The Intuitive Way of Knowing

The Intuitive Way
of Knowing

A Tribute to Brian Goodwin

Edited by David Lambert,
Chris Chetland,
and Craig Millar

Floris Books

Published by Floris Books in 2013

© 2013 David Lambert and Chris Chetland
All rights reserved. No part of this book may be reproduced without the prior permission of Floris Books, 15 Harrison Gardens, Edinburgh
www.florisbooks.co.uk

 This book is also available as an eBook

British Library CIP Data available
ISBN: 978-086315-965-7
Printed in Great Britain
by Bell & Bain Ltd, Glasgow

Contents

Editors' Preface · 7

PART 1. PERSONAL MEMORIES

Brian Goodwin, The Magical Black Knight · 11
Stuart Kauffman

Brian Goodwin · 15
Lewis Wolpert

Brian Goodwin at Sussex in the 1970s: personal reminiscences · 17
Claudio D. Stern

A Long Friendship · 35
Frederick W. Cummings

An Interview with Brian Goodwin: 1 · 37
Stephan Harding

PART 2. EVOLUTIONARY BIOLOGY AND PHILOSOPHY

Biology without Darwinian Spectacles · 45
Brian Goodwin

Form and Function in Biology: Placing Brian Goodwin · 55
Michael Ruse

Epigenetics & Generative Dynamics: How Development
  Directs Evolution   71
    Mae-Wan Ho

Darwinian Spectacles and other 'Ways of Seeing' Evolution   96
    Craig Millar and David Lambert

From Goethe to Goodwin, via von Foerster   108
    Margaret Boden

An Interview with Brian Goodwin: 2   120
    Stephan Harding

PART 3. THEORETICAL BIOLOGY

Complexity and Life   127
    Fritjof Capra

The Language of Living Processes   141
    Philip Franses

Keeping the Gene in its Place   153
    Johannes Jaeger and Nick Monk

An Interview with Brian Goodwin: 3   187
    Stephan Harding

References   193

Brian C. Goodwin: Selected works   215

Index   220

Editors' preface

It was obvious to anyone who knew Brian Goodwin (1931–2009) that he was a quite remarkable man. He was not only extremely articulate but he thought about biological systems in a way that few people did. His skills in communication not only endeared him to those biologists who sought a new way of looking at evolutionary biology in particular, but even those who felt that the current perspective was essentially adequate.

Brian had a mix of strong mathematics such that as his close friend John Maynard Smith expressed it: 'He always got his sums right'. His background in experimental biology also gained him respect among developmental biologists, but it was his theoretical work that meant that Brian Goodwin stood out, in even the most distinguished of crowds.

This volume comprises papers by his friends and colleagues, dedicated to his enduring memory. Just as Brian had a diverse set of skills, backgrounds and interests, so too does this book. In the opening section, there are personal reminiscences from his close friends: Stuart Kauffman, Lewis Wolpert and Claudio D. Stern. Fred Cummings also writes about their long friendship and common interests. There follows the first of three parts of Stephan Harding's interview with Brian that took place shortly before his death.

The middle section is prefaced by one of Brian's most distinctive papers 'Biology without Darwinian Spectacles'. This section centres on evolutionary biology and its philosophical underpinnings and includes writings of authors with such diverse viewpoints as Michael Ruse and Mae-Wan Ho, Craig Millar and David Lambert. Maggie Bowden explores a theme of central importance to Brian Goodwin, namely the relationship between Goodwin's work and the intellectual debt he acknowledged to Johann Wolfgang von Goethe. The section closes with elements of the interview by Stephan Harding relating to evolutionary biology and the perspectives that underpin it.

The final section of the book has a focus on more general theoretical issues, reflecting the fact that Brian Goodwin's breadth of intellect shone through in many other areas of science. Fritjof Capra begins this dialogue with a paper on 'Complexity and Life'; Philip Franses writes on 'The Language of Living Processes'; Johannes Jaeger and Nick Monk on 'Keeping the Gene in its Place' – all exploring issues that were of particular interest to Brian. The editors hope that all biologists and others with an open mind, keen to explore the mental excitement of novel approaches to theoretical and evolutionary biology, will similarly want to explore this collection of papers which are warmly dedicated to a most remarkable and lovely man.

David Lambert, Brisbane, Australia
Chris Chetland, Auckland, New Zealand
Craig Millar, Auckland, New Zealand

PART 1

PERSONAL MEMORIES

Brian Goodwin: The Magical Black Knight

STUART KAUFFMAN

Stuart A. Kauffman is a theoretical biologist with an interest in the origin of life and molecular self-organization. He has been a professor at the University of Calgary in biological sciences and physics and astronomy, among other appointments, and worked in complexity studies with Brian Goodwin and others at the Santa Fe Institute, New Mexico. His most recent book is Reinventing the Sacred: A New View of Science, Reason, and Religion.

Brian Goodwin, the Black Knight, Poet of Biology, visionary, brilliant, a lovely human being, was my best friend, almost a brother.

I too have been lucky to have a long career. Brian, I have learned, was always a decade ahead of me, and virtually always right. I thought, each time, my dear friend was nuts.

I think, in Brian's passion for a Structuralist Biology, in which he, in the tradition of D'Arcy Thompson, was a leader, Brian's only rather stubborn mistake was to underestimate natural selection. How many arguments we had, I too seeking to meld Darwin into some larger framework of laws of self-organisation *and* selection.

In the end, Brian too admitted Darwin into the Goodwin canon.

But first, just images of Bri. My wife and I live in Santa Fe. Our first house lay on a sweet meadow with views up the Jemez mountains to the west and Sangre de Cristo, the southern tail of the Rockies, to the east. Lightning and thunder would roll in from the Jemez, vast sheets of sheet lightning. What did Brian do? He yanked a metal poled umbrella and ran umbrella upheld to the threatening clouds towards the lightning, laughing madhatterly.

Henfield beer, corn, kids, and a quiet walk, just the two of us, Brian cleanly explaining to me the rudiments of structuralism in biology. I was eight years his junior. I learned from my best friend.

First encounter: I had worked on random Boolean networks as models of genetic regulatory networks, from age 24 as I entered medical school. In a bookstore in San Francisco I stumbled upon his thesis, *Temporal Organization in Cells,* done under C.H. Waddington in Edinburgh.

Bless Brian, he was the first among us to try for a dynamical theory of the integrated behaviour of the genetic regulatory system, in the first years following the discovery by Jacob and Monod that genes could inhibit or activate the transcription of other genes. In doing so, Brian, past Rhodes Scholar in mathematics, with a degree from McGill, foresaw the core of contemporary Systems Biology in his PhD thesis.

How much can one ask of a young scientist forty years ahead of his time?

Temporal Organization in Cells, later a gift to me inscribed with Brian's heart: 'There is no truth beyond magic', of course terrified young me. Had this man done *everything* I was trying to do and beaten me? Good God!

Almost all young scientists go through this experience.

But no, Brian's approach was lovely but not my random Boolean network ensemble approach. Thank goodness, I'd live another day, barely.

We met magically. Warren McCulloch, early cybernetician at MIT had invited Brian and his first wife, Pearl, to live with him in Cambridge. Warren had invented with Walter Pitts, the idea of formal neurons which, in arbitrary feed forward arrays, could, with arbitrary logic, compute any desired Boolean function of the on/off states of the input layer, leading to a 'Calculus of the Ideas Immanent in the Mind', a 1943 paper which set the stage for neural network theory as known today.

Warren and his wife Rook were lovely. For reasons never quite made clear to me, Brian and Pearl would climb out the window of their bedroom to explore Cambridge. I don't know why, the front door worked, and Warren and Rook were, well, spectacular, open and generous. Maybe Brian and Pearl liked skinning their shins?

Some eight years later, I was lucky enough with my new wife, Elizabeth, while in third year medical school, UCSF, to be invited by Warren to live and work with him for three months at MIT.

'Stuart', he said with unbounded enthusiasm one day, 'come meet Brian Goodwin!'

I think I immediately fell in love with the then 35 year old man who became a cornerstone of my life: tall, whimsical, full of ideas, utterly open, able to listen and listen to me babble on about my dearest ideas which would, of course, transform all, and listen intently, and respond with ever encouragement. How Brian shaped my growth into a young scientist. I want to cry now as I remember. How, over the years, we played almost symmetrical roles, each listening intently to the other's dreams, work, hopes.

Listening just recently to Brian give a Schumacher lecture as his wife Christel visited us just weeks ago, I sensed again the magic, gentleness and wisdom of my now old friend, with metastatic prostatic cancer, riding his bike and walking the dogs.

We have lost a unique voice. I still love the man and mind.

And his science? While still at the University of Sussex, Brian built a model of the complex development of a single-celled organism, *Acetabularia*, modelling calcium diffusion and a complex local interaction with the cytoskeleton that predicted in astonishing detail the formation of the complex mature organism, with whorls of filaments forming at one end. It was a triumph. The film is available. This was Brian's Structuralism on the march, morphogenesis beyond the scope of genes alone.

The response of the biological community?

Silence. We were in the midst of, and just now emerging from, the genocentric view of biology. We forget, but Brian knew, that genes only determine the time and location where gene products are made and what those products are. The rest is up to physics and chemistry, atoms to cells, Brian's Structuralism. Again Brian was ahead of biology by decades.

He was ignored. It is vital to keep my dear friend's work alive and known, for it will help lead just the transformation in biology of which he dreamed.

A Science of Qualities. I thought, as usual, my dear friend was nuts. But he was right and I was wrong. Darwin's *Origin of Species* has not a single equation. Darwinian preadaptations are beyond prestatement, and, I claim in parallel a decade after Brian, are not mathematisable. They are qualitative novel functional features of organisms, and there is no finite statement of possible functionalities in indefinitely diverse selective environments. Beyond Structuralism, and embracing Darwin, the biosphere's becoming is a matter of qualities, novel functionalities by which organisms come to make livings with one another. We cannot prestate the becoming of niches. More, the emergence of a new species, or new organ, say a swim bladder from the lungs of lung fish by Darwinian preadaptation is not only an un-prestatable new functionality, it creates a new 'adjacent possible' niche, for a bacterium might evolve that lives only in swim bladders, as mycoplasma lives only in the lungs of sheep. So, magically, and unprestatably, the biosphere's evolution literally creates the empty adjacent possible niches that it then fills. The biosphere creates the non-random, context dependent, qualitative possibilities that it will become! So does the economy in its evolution.

The paragraph above is my own view, but I think Brian would have agreed. It is part of the foundation for the need for a science of qualities, where Brian foresaw that not all in the becoming of the biosphere, let alone economy or history is mathematisable and deducible as theorems entailed by a prestated set of axioms.

The world is richer, Horatio, than all our dreams. But Brian saw this. Again he was forty years ahead of his time. Forty years from now, we will begin to embrace the idea that we cannot have mathematisable laws for all that unfolds in the becoming of the biosphere, economy, history. We will ask: When can we have law? When not? Why? How do we live, not knowing what *can* happen, where reason is an insufficient guide to living our lives forward. Can we have a 'science of qualities'? In what sense of 'science'? Some kind of understanding is called for, but it seems we cannot prestate the becoming of the world.

Brian's intuition that we need some kind of 'science of qualities' once again was his deep visionary intuition. He was right. Again.

I want to close on a deeply personal note about my dearest friend. Brian always feared he could not love. But in Christel his heart opened, and with her, he completed the arc of his life. How blessed they both, and he, the Black Knight, home again as the arrow of his life pierced the soil and he left us.

Brian Goodwin

LEWIS WOLPERT

Lewis Wolpert is an emeritus professor at University College London. He is a developental biologist and this was the topic that he often discussed with Brian – they had very different views.

Brian was a very good friend even though we had some basic differences about biological mechanisms. He was a Canadian and first studied plant biology but then became a mathematician at Oxford University. In the early 60s we met and had a small group to discuss biology that included Stuart Kauffman and Anthony Robertson. We always disagreed as Brian was not interested in cells as the basic units, and took a more general systems approach.

His first contribution was to show that organisms are essentially rhythmic systems accounting for the universality of biological clocks. He was a scientist of outstanding calibre who then went on to develop an alternative to the neo-Darwinist notion that natural selection acting on randomly mutating genes is the fundamental process that drives evolution. He argued that with a focus on genes and how they change in time, there's a tendency to ignore the spatial dimension of organisms. To describe their spatial patterns in organisms, he believed you needed field theories like those used in physics to explain spatial order. Goodwin was a visionary in thinking of the whole genome as one dynamical system, and became a strong advocate for 'structuralism' in biology, the thesis that it is not true that all of the order in biology is due to natural selection alone. He was anti-reductionist. He made a highly influential mathematical model that elegantly simulated whorl formation in *Acetabularia*, a tiny unicellular marine alga shaped liked a miniature umbrella.

When he was a Reader in Biology at Sussex University I used to visit Brian and we had debates in front of the students. I used to call his ideas mystical and that they floated in the clouds, whereas he said that I was a boring reductionist. He was determined to deal with the embryo as a whole rather than a collection of cells.

I even wrote one paper with Brian together with several others including John Maynard Smith in 1985 on the constraints that development put on evolution.

Later he became Professor at the Open University and ended up at Schumacher College. His best known book is *How the Leopard Changed its Spots: The Evolution of Complexity*. He became hugely influential in theoretical biology, and was very much admired. I even envied his good looks.

We used to meet regularly and we used to take our tennis things to meetings and play when we could. His service was excellent. Some time ago together with Brian at a meeting in Berlin, I was rather down and told Brian that I was worried about dying. Later that evening my phone in the hotel rang and it was Brian. He told me to stop worrying, as death was the next great adventure. It really helped me – Brian was so kind.

I did visit Brian after he had his heart problems, from which his recovery was impressive. We had lunch together about six months before he died. He and his wife were planning to go to the University of Stellenbosch in South Africa where Brian had been invited, so that he could write a book. As a South African I was planning to put him in contact with friends of mine who lived in Capetown near the University. I miss him, as do many others.

Brian Goodwin at Sussex in the 1970s: personal reminiscences

CLAUDIO D. STERN

Claudio D. Stern works in the Department of Cell & Developmental Biology, University College London, and did both undergraduate and postgraduate studies under Brian Goodwin at the University of Sussex.

Biology at Sussex

The School of Biological Sciences at Sussex was an extraordinary environment in the 1970s. Many people agree that most of this was due to the vision of John Maynard Smith, who started it. Among the stories I was told then (by John Maynard Smith himself) was that, from his deathbed in 1964, J.B.S. Haldane, the great evolutionary biologist, had proposed to John the challenge of founding a Biology department at the then new University of Sussex. John's subsequent choice of recruits was to lay the foundations for a new, vibrant intellectual environment. Among the first recruits in 1965 was J.H. (Jimmy) Sang from Edinburgh, a geneticist interested in *Drosophila* as well as poultry. Jimmy and John were to alternate as Dean of the new School through the 1970s. The same year, John recruited Brian Goodwin. Brian had studied Biology at McGill, Canada, then Mathematics as a Rhodes Scholar at Oxford, then undertook a PhD in Edinburgh under the great developmental biologist, geneticist and polymath C.H. Waddington, before going to MIT as a postdoc. Brian told that J.B.S. Haldane had read his recent book *Temporal Organization in Cells* (Goodwin 1963), and although he claimed not to understand any of the maths, was impressed enough to suggest to John Maynard Smith (who also claimed not to understand the maths) that he should consider Brian for a Readership in Embryology in the new Department. Brian and John would talk about key questions

17

in Biology regularly ever after, although their fundamental view of evolution could not have been more different.

Sussex felt like the centre of the world for Developmental Biology in the 1970s. Jimmy Sang had built up a group of *Drosophila* geneticists on the first floor (Jimmy Sang, Robert Whittle, Janet Collett, whose husband Tom was a neurobiologist also in the department, along with others who were also interested in evolutionary questions: Deborah and Brian Charlesworth). This was a group of scientifically rigorous *Drosophila* geneticists; they were all deep thinkers and were constantly engaged in discussions with each other. They loved experimental science and were also superb, committed teachers. Jimmy Sang stopped working on poultry and devoted himself entirely to the fly. He pioneered the cell biology of fly development by exploiting a combination of quite different techniques – mutants that affect cell size or other aspects of cell behaviour, endocrinology, biochemistry, cell manipulation and some of the first methods for insect cell culture. Among his early PhD students were Peter Bryant and Jonathan Cooke (both working on tumour suppressor genes in the fly and the influence of steroids on tumour formation; more about both of them below) and a little later Mary Bownes (pioneering different methods for cell manipulation in *Drosophila*; she is now Vice-Principal of Edinburgh University). Robert Whittle had been recruited by Jimmy Sang in 1970 to fill the vacant lectureship created by the departure of Ben Lewin (who had then been working on the first version of his influential book, *Genes* and upon leaving Sussex went on to found the highly successful journal *Cell*). Rob became interested in the extraordinarily difficult problem of how cells construct patterns during development, integrating signalling, growth and differentiation. I feel that his scientific contributions have not been given the recognition they deserve, except perhaps through the work of his postgraduate student Phil Ingham, who subsequently made hugely important discoveries on Hedgehog signalling in both fly and vertebrate development. Rob was also a superb tutor; when I arrived as an undergraduate in 1972 he was assigned as my personal tutor and my first years at Sussex would have been much more difficult without his guidance.

There was another group of academics with more diverse interests centred on the fifth floor (Brian Goodwin, Chris Ford and Gerry Webster) and several others scattered among various parts of the building who were mainly molecular biologists (a newly-emerging discipline in those days) but were studying problems of direct developmental interest, including Robin Cole (cell differentiation including myogenesis and haematopoiesis), Alan Lehmann (DNA repair), Sydney Shall (poly-adenylation and gene expression) and Neville Symonds (bacteriophage biology). There was also, of course, an extremely strong group of evolutionary biologists led by John Maynard Smith which included Paul Harvey and others. They shared the fifth floor with Brian Goodwin. Others that passed through around that time included future luminaries like Paul Nurse (as MRC senior research fellow; later to win the Nobel Prize for elucidating the cell division cycle and now President of the Royal Society), Brigid Hogan (as a lecturer from 1970–74, later to make many important contributions to understanding the biology of Hox genes and many aspects of organogenesis) and Georgina Mace (with whom I shared a house on Egremont Place, Kemp Town in 1977, and who was working on the fifth floor with the ecologists and evolutionary biologists and is now an influential leader in conservation biology). Among the early graduates was also Adrian Bird (later to become a leader in the study of the control of gene expression by DNA methylation).

Surprisingly, the different groups of developmental biologists were fragmented intellectually and philosophically as well as physically within the building. But it was a privilege for those of us who were undergraduates there at this time to be taught by all of these inspiring people. Having been an undergraduate in the School also helped me later, as postgraduate student, to continue to talk with everyone in the staff tea room, which was adjacent to our corridor on the fifth floor and frequented by most staff, and with John Maynard Smith on the same floor, who was interested in, and always seemed to have thought of, everything. There were frequent seminars by invited speakers, which included many distinguished sabbatical visitors. The seminars were often quite provocative and I remember some very vividly. For example, Compartments (lineage-delimited fields of cells in developing systems) were just being discovered by Antonio García-Bellido, Ginés Morata and Peter Lawrence in *Drosophila* (Morata and Lawrence 1975) and all of these came to visit, especially to interact with Jimmy Sang and Rob Whittle. We started to discover these seminars from our second year as undergraduates and it was customary

19

for us to attend, and to follow the seminars by extended discussions about how development works, or anything else.

The atmosphere at the time was always one of debate and discussion. It started with politics. Everything seemed grey. We, the students, frequently occupied the administration building (Sussex House) to express our discontent about poverty, the imprisonment of Nelson Mandela, Apartheid and all other forms of inequality, nuclear weapons and government interference. A major victory was achieved when we successfully prevented Hans Eysenck from entering the Sussex campus to give a lecture to present his theory that whites were the more intelligent race – extraordinary views for someone who had left Germany because of his differences with the Nazi regime. There were almost nightly and frequently quite wild parties on and off campus, but somehow none of this seemed to interfere with the pleasure of discussing burning scientific and philosophical questions deep into the night. Everyone was involved – undergraduate and postgraduate students, academic staff and even some of the cleaning staff, who were part of the family. And somehow there was even time left for other things – sports for some, the arts for others, or rambling along the Sussex Downs (the stunning white chalk cliffs along the South coast).

Among Brian's colleagues on the fifth floor was Gerry Webster, with whom Brian interacted regularly (Webster & Goodwin 1996). He was one of the Literati – a deep thinker with very broad interests spanning History and Philosophy of Science, Mathematical Biology (both of which he taught, with passion but also extraordinary density) and Developmental Biology, especially regeneration. He had worked with Lewis Wolpert on regeneration of the tentacles in *Hydra*, which can regenerate in the absence of cell division ('morphallaxis'). Gerry only supervised DPhil students very occasionally, but around this time he did have a particularly good one, Vernon French, who was doing remarkable experiments, working in one of a series of temporary structures known as 'the terrapins' next to the Biology building. Vernon was using the cockroach limb to study regeneration. Cockroach limbs, unlike *Hydra*, require cell division for regeneration ('epimorphosis'). Vernon conducted a series of systematic rotation and recombination experiments and discovered that unexpectedly, if part of a cockroach leg is amputated and then rotated 180° before being pasted back to the stump, three legs regenerate rather than one. Similar results are obtained with left-to-right transplants without rotation. These studies (French 1976, 1978, 1980) comprise a real *tour-de-force* of experimental embryology and led to a hugely influential model for limb regeneration and pattern formation: the polar coordinate model (Bryant *et al.* 1977; French 1976; French *et al.* 1976). The authors of the model were Vernon, Peter Bryant (a former student of Jimmy Sang's) and Susan Bryant, Peter's American wife who had done her PhD in London at St Mary's Hospital Medical School with Angus Bellairs (a famous herpetologist and husband of Ruth Bellairs, the embryologist from University College London). Each of them contributed results from a different experimental system: Vernon from his cockroach limb transplantations, Susan from initial observations in lizards made during her PhD and subsequently in the newt *Notophthalmus,* and Peter from fly imaginal discs, following on from his work at Sussex. Regeneration in all three systems seemed to be governed by just two simple rules: intercalation (missing regions were regenerated by 'averaging' with their neighbours) and distal transformation (completion of the structure from the level of amputation down to the tip). The model was a masterpiece of biological inference and abstraction. It immediately became one of the most exciting new ideas in Biology and was a topic of daily, heated discussions in the tea room. The Bryants then went to the University of California at Irvine where Peter became the head of Developmental Biology as well as chief editor of the homonymous journal. They both continued to do superb work on pattern formation, working on fly and vertebrates. Upon Vernon's departure I was given his terrapin office in 1975, which I was to share for the best part of the next two years with a colony of cockroaches that had refused to accompany Vernon to Edinburgh.

The other academic member of the group of developmental biologists on the fifth floor was a young lecturer, Chris Ford. He was lumbered with much of the undergraduate teaching and I remember him as an extremely clear and didactic lecturer, who made developmental biology immediately accessible. Being only a few years older than the students he taught, he was especially approachable. Chris used mainly biochemical methods and was studying oocyte maturation and control of the cell cycle in early *Xenopus* embryos. Chris was very knowledgeable about embryology and was almost an 'honorary' member of the Goodwin/Webster group of developmentally-minded biologists on the fifth floor. However, his way of thinking was much more molecular and I think that his work was effectively ignored by these colleagues, as well as by the *Drosophila* geneticists. Perhaps for this reason his collaborations with other biochemists in the School, like Sydney Shall, were more productive. During my time at Sussex Chris had a technician, Dave Brill, and a PhD student, Christina Smith – they were endeavouring to discover universal controllers of DNA replication and of the cell cycle (Smith *et al.* 1980; Smith *et al.* 1983).

Brian's personality

Everyone agreed that Brian was inspirational. But I still find it very difficult to determine why, or in what way. Students adored his lectures, although many were very difficult to understand because Science was often wrapped up with esoteric thoughts, or concepts which he only partially explained. It was obvious that there was always something more behind what he said, which he purposely kept back. With time I discovered that some of his arguments did not stand up to close scrutiny or prolonged discussion, but it took quite a lot of persistence as well as knowledge to do this. Most of the time, he would produce something new and different out of the hat, for which the interlocutor was not prepared. You would leave the discussion in awe of the breadth of his knowledge and even if not always convinced, at least mystified. There were also unfathomable aspects of Brian's interests and personality – around 1972/3 he took a sabbatical to travel to Mexico in search of Carlos Castaneda's possibly mythical Don Juan. He did not reveal to us whether he ever found him, but his dress sense on return was different, and his shaman-like aura seemed to have grown. Perhaps he had gone in search not only of knowledge and enlightenment but also seeking a cultural link to his own North American roots. Throughout his life he was also attracted to Indian mysticism and spiritualism – joining Maharishi Mahesh Yogi's Transcendental Meditation group was not a difficult leap for a scientist because of the pragmatic approach of this group, but some of his other beliefs were more difficult to reconcile with Science. These were all indications that Brian was after some form of spiritual enrichment.

Scientifically, Brian also appeared to be on a quest, perhaps similar to that of his PhD supervisor C.H. Waddington, who sought a 'universal theory of everything'. Waddington looked at genetics, cell biology, embryology, mathematics and evolution as well as reading and writing about politics and philosophy. Brian's work with Waddington had explored several problems especially from a theoretical point of view but did not lead to any publication. However it is obvious that their discussions were fundamental to Brian's development as a scientist and thinker, and both of them regularly attended the meetings of theoretical biologists organised by Waddington in Villa Serbelloni, in Northern Italy. The inside cover of a volume of the proceedings of the first of these meetings (Waddington 1968) includes a photograph with both of them together with other Sussex figures of the time such as experimental psychologist Christopher Longuet-Higgins and John Maynard Smith.

Some of us who passed through Brian's lab always aspired to combine theory and experiment. Some even inherited Brian's and Waddington's desire to define a 'universal theory of everything'. We were jealous of our physicist colleagues who could write $E=mc^2$ and be satisfied with the predictive power of something so simple and beautiful, never driven to find the material basis for why this is so. Just as I was arriving in Brian's lab in 1975 to start my DPhil, I had arranged to visit Waddington in Edinburgh on October 2. I don't think I knew what I wanted from him – just to meet him, perhaps expecting Wad to exude even more inspiration than Brian. As I left the Biology building with a small bag, heading for the train station, I ran into Brian who asked me where I was going. I told him. He said, 'Haven't you heard? Wad died last week'.

Unlike Waddington or John Maynard Smith, or almost anyone else, Brian believed that genes could not possibly be at the centre of Biology – including development and evolution. He disliked the idea of predetermination and that anything might be fixed without the possibility that the individual could change it through his or her actions. It may be this, more than any specific scientific argument, which determined his rather extreme anti-gene stand, for which he was widely known throughout his career.

If not genes, what? Where is the information that governs development and evolution encoded? Brian believed that this information is carried by the Laws of Physics and Chemistry and that biological systems are governed by system-wide properties. In the right environment, elements can self-assemble – therefore the key for a developing organism and its cells is to generate the right environment for this to happen. He was also particularly interested in the dynamics of physical phenomena and realised that *time* is very rich in information content. This caused him to explore, theoretically, whether waves and oscillations of various components could convey spatial and temporal information to cells to instruct them how to behave during development. He proposed a particularly important model: the 'phase shift model', in collaboration with theoretical physicist Morrell H. Cohen from Chicago (Goodwin and Cohen 1969), suggesting that if two or more oscillators of different wavelength operate within a tissue, cells could obtain spatial and temporal information by measuring the phase shift between the two frequencies as they propagate through the tissue. Jeremy Brockes recalls that this model attracted the interest of Francis Crick who at the time was thinking about diffusion and other physico-chemical processes in relation to positional information.

Brian sometimes presented his views as much more extreme than they really were, to provoke thought and scepticism. This iconoclastic approach partly accounts for Brian's influence on those around him but the same also caused him to be generally dismissed by much of the establishment. Molecular biology was just being born, and many were deaf to the ideas of someone who constantly announced that genes are unimportant. Had Brian placed gene action alongside the dynamic principles and physics of matter, rather than as mutually-exclusive alternatives, his lasting influence and contributions might have been very much greater.

Brian's lab

Brian remained a theoretician and thinker throughout his academic career at Sussex, but like Waddington, he believed that it was extremely important both to observe the living system and to experiment with it to understand the laws that govern its behaviour. It is for this reason that he ran a lab, and quite a large and diverse one. He did not engage in experiments himself but was amenable to looking down a microscope if it stimulated discussion about something in which he was interested. Although he was rarely receptive to information or biological examples that did not obviously help address the problems of current interest to him, he was always willing to discuss deeper issues and ideas with those who carried out the experiments in his lab.

One of the early (1969–72) postdocs to work with Brian was Jonathan Cooke, who had completed a PhD with Jimmy Sang. At that time Brian's space was on the first floor, closer to the *Drosophila* developmental geneticists. Jonathan and Brian applied for a grant from the Science Research Council (SRC; now BBSRC), to repeat and extend Hans Spemann's famous organiser experiments in the amphibian *Xenopus laevis*, to determine in detail the cell biological properties of the organiser, whether the embryo can compensate for its removal, the spatial and temporal characteristics of the signals and of the embryo's responses to them. Five classic papers resulted from this work (Cooke 1972a, b, c, 1973a, b). However, even with the background of having done his PhD with the developmental geneticists and then a postdoc with Brian, Jonathan was unable to get the two to engage constructively in discussions about developmental mechanisms. Jonathan himself remembers: 'I wanted to explore "the other side" in the overarching gene/anti-gene debate that I'd been exposed to in the coffee room. Jimmy took several years to recover in our personal relationship from the sense of intellectual betrayal he clearly felt [when I joined Brian as a postdoc]'.

During this time Jonathan also looked for dynamic phenomena, filming frog and chick embryos as well as a hydroid, *Tubularia*, the latter in collaboration with Gerry Webster and during a trip to Naples where he overlapped with both Brian and Waddington, who were resting there after one of the meetings at Villa Serbelloni. He observed optical waves as well as oscillations of cell behaviours with periods of about 100 seconds, which attracted Brian's interest. Jonathan recalls that these observations sowed a seed that was later to resurface as part of his 'clock and wavefront model' for somite formation (Cooke & Zeeman 1976), which he proposed to account for the abnormalities observed after exposing *Xenopus* embryos to a short heat shock (Cooke 1978; Cooke & Elsdale 1980; Elsdale *et al.* 1976).

Brian moved to the fifth floor around 1973. By then he had a mix of students, postdocs and technicians who worked on an unusually wide array of model organisms. Most of these people had their own small lab scattered along the fifth floor corridor and some of them even had their own technician. The research was funded by a few research grants and an allocation from the School to each academic. There was also a stream of distinguished and often very stimulating visitors, such as theoreticians Art Winfree (interested in modelling oscillators) and Stuart Kauffman, who was modelling gene expression control and attempting to build gene regulatory networks, at that time to understand the phenomenon of transdetermination. Stuart Kauffmann later became a regular visitor and was one of very few people with whom Brian could communicate very well on scientific issues.

In the middle of the corridor, opposite Brian's office, was a small lab used by an advanced PhD student, Lür Willnecker. Lür was studying the development of the pattern of spots in the wing of the moth, *Ephestia kuhniella*. At the time, work by Peter Lawrence and others on several insect species had led to the proposal that a wave of positional information travels along the prospective cuticle of the insect and that the dynamics of this wave determines the pattern of stripes, eye spots and other pigmented features. This was of natural interest to Brian because it represented a tractable example of a global, dynamic event with informational content. There was no information about the nature of the wave, but more importantly its physical attributes (speed, interaction with other waves, etc.) were open to investigation. Lür's project consisted of using heat shock, delivered at precisely controlled times, to try to disrupt the passage of the wave and thus determine its speed and establish a connection between physical properties and the resulting pattern. Lür had obtained some interesting results and had written up and submitted his DPhil thesis around 1975. The appointed examiners were to be John Maynard Smith as internal and Jonathan Cooke as external. About a week before the oral examination, the examiners decided to call it off because they found the thesis too difficult to understand. Lür was asked to re-write it. However, at this time Lür was being lured by the teachings of Bhagwan Sri Rajneesh and decided to travel to Poona to meet Bhagwan, in search of enlightenment. On return to Sussex a few months later, Lür told Brian that Bhagwan had asked him to throw the thesis into the Ganges. Brian said: 'Did you?' to which Lür replied that he had. It took considerable further questioning before Lür admitted that he had kept a copy. However, that was the last time any of us saw Lür as he never reappeared and never submitted his amended thesis or published his findings.

At the end of the corridor, turning left, was a very small air-conditioned room that allowed the lights to be controlled by a timer and also had tanks of recirculating water. Here is where Stelios Pateromichelakis studied *Acetabularia*, a unicellular plant capable of regenerating its large cap, assisted by Susan Kirk-Bell. The attraction of this system for Brian was obvious, because the cap can regenerate fully even when the foot (which contains the nucleus and therefore the entire genome) is removed: regeneration can occur in the absence of genes. Stelios looked for non-genetic cues that might encode the information for patterning the regenerating cell and focused his attention mainly on electrical currents and Calcium fluxes. Since this is an enormous cell (2–4 cm long) it is possible to lay it out along a small plastic box subdivided into compartments, allowing electrical measurements and application of different ionic solutions to the different chambers, or imposing an artificial electrical gradient along its length. The main finding was that Calcium levels are important for regeneration of the cap but not for growth (Goodwin & Pateromichelakis 1979). *Acetabularia* remained a topic of great interest for Brian throughout his scientific life and even formed the basis for the cover of his 1976 book, *Analytical Physiology of Cells and Developing Organisms* (Goodwin 1976). Several years later Brian took this a step further in collaboration with Paul O'Shea using a vibrating probe – they succeeded in measuring large extracellular currents centred around the regenerating cap (O'Shea *et al.* 1990). At the time that Stelios was conducting his experiments in Brian's lab, Lynn Trainor, a mathematical biologist from the USA was spending his sabbatical at Sussex, working with Brian. Lynn would become a regular visitor whose work with Brian included further modelling of *Acetabularia* regeneration, suggesting that in systems that undergo changes in Calcium, mechanical strain fields could interact with these to provide the spatial information to pattern the regenerating cap (Goodwin & Trainor 1985).

Other regenerating systems were studied in the lab as examples of systems that generate complexity from comparable simplicity. Malcolm Maden was a postdoc studying limb regeneration in the axolotl (Mexican newt) limb. He was more senior and more experienced than the others, and had also secured research grants from MRC and SRC for his work. He also had his own technician, Wendy Neilson. A natural experimentalist, Malcolm arrived having published a number of carefully executed and important papers with his PhD supervisor Hugh Wallace, describing in detail the formation and cell biology of the blastema, a region of dividing cells that forms at the stump after amputation and which is the source of cells for the regeneration. In Brian's lab, Malcolm's initial project was to investigate whether limb regeneration in the axolotl might involve signalling from cellular centres similar to those that had been discovered in the embryonic chick limb bud: the zone of polarising activity (ZPA), apical ectodermal ridge (AER) and progress zone. The main conclusion was that limb regeneration in the axolotl was not exactly comparable to embryonic limb development (Maden & Goodwin 1980). During these studies Malcolm acquired his own DPhil student, Neil Turner. The polar coordinate model was being formulated by Vernon French, who had just finished his DPhil with Gerry Webster (see above), and this was therefore a topic of almost daily discussions among all of us, as we speculated whether a similar system might govern development and regeneration of each of our systems. Neil's project was therefore to test whether it could apply to regeneration of the axolotl limb. Indeed, when the blastema (growth zone) at the stump of an amputated axolotl limb was rotated by 180°, three supernumerary limbs regenerate, just like Vernon had found in the cockroach. This finding attracted huge interest and was published in *Nature* (Maden & Turner 1978). Even though this was done in his lab, Brian did not feel that his intellectual input had been sufficient to justify being included as an author (as he had done with Jonathan Cooke's work some years earlier). However I now believe that another factor might have been Brian's dislike for models based on fixed coordinate values ('positional information') including that proposed a decade earlier by Lewis Wolpert (Wolpert 1969), and therefore did not want to be associated with this type of mechanism. He strongly favoured the view that spatial information was global and dynamic.

Neil's thesis work was not without problems but did lead to a second paper (Turner 1981) in which he proposed that the frequencies of supernumerary limb formation in blastema rotation and other experiments did not exactly fit the polar coordinate model. Neil suggested a probabilistic model for regeneration, which did not catch on. Neither Brian nor Malcolm were co-authors, partly reflecting that Neil's interpretation was not shared by either and perhaps also revealing that Malcolm had adopted this attitude from Brian's reaction to his earlier paper. Malcolm Maden then left Sussex for the National Institute of Medical Research at Mill Hill where he continued his limb regeneration studies. Among his many other important discoveries was the finding that retinoid (Vitamin A) signalling is crucial for cellular memory of proximo-distal positional information in the limb: when amputated axolotls are reared in a medium with high Vitamin A content, they forget the level at which the amputation had been made and regenerate a new limb from the shoulder (Maden 1983). Malcolm went on to become a world authority on retinoid signalling during development, regeneration and disease and is now a Professor at the University of Florida.

A small molecule, cyclic-AMP, was very trendy in the early 1970s. It seemed to be involved in everything and ideas abounded about whether it might control various aspects of development, including its possible role as a carrier of spatial and temporal information. In my final year as undergraduate (1974–75) I was looking for a project on which to write a dissertation and after discussing various topics with Brian, who was to supervise me, I decided to survey these many roles of cyclic-AMP. I wrote a long dissertation with hundreds of references, and including a complex model (something like what some call an 'interactome' today) with cyclic-AMP at the centre. Brian was not very interested. However he was interested by one particular observation involving cyclic-AMP: its role as a chemoattractant in *Dictyostelium*, the cellular slime mould. This simple organism normally lives as single cells which divide asexually; when stressed by the shortage of food it changes strategy: the individual cells aggregate to form a multicellular slug which migrates and then settles, making a fungus-like structure with a foot and a long stalk at the top of which is a fruiting body whose cells undergo meiosis and become germ cells that scatter to undergo sexual reproduction. What was of particular interest to Brian in this context is that cyclic-AMP is released in a pulsatile manner which, together with a refractory period of cell responses to it, ensures that it can encode directional information to the migrating single amoebae. This was therefore an ideal system for Brian to explore the role of dynamic signals in imparting order and complexity in a developing system.

Cornelis (Kees) Weijer was an MSc student from the Hubrecht Laboratory in Utrecht, where he had been working with Astrid

Lindenmeyer and Pieter Nieuwkoop on signalling during amphibian development. In Utrecht Kees had also been working on *Dictyostelium* with Tony Durston who was in communication with Brian concerning the possible roles of cyclic-AMP oscillations in providing developmental information. Brian had also visited the Hubrecht Laboratory in 1974 for a symposium for theoreticians and experimentalists, which persuaded Kees to move to Sussex to work with Brian. Kees's experimental project was meant to be on *Acetabularia*, studying the dynamics of the propagating Calcium waves. However he also introduced the slime mould to the lab and continued some of the experiments he had started with Tony Durston, examining the kinetics of cell aggregation and the mechanisms that apportion the correct number of cells to each cell type in the slug stage. Kees stayed at Sussex for eighteen months, then returning to Tony Durston for a PhD before taking an independent position in Munich, after which he eventually became Professor and Head of the Division of Life Sciences in Dundee.

When I arrived to begin my DPhil in the late summer of 1975, Brian asked me to choose a topic and experimental system. I asked what was on offer, to which he replied: 'Well, you can have *Acetabularia*, amphibian limb regeneration, frog somite formation, or the chick'. I asked what he meant by 'the chick'; his response was that anything was possible, but maybe gastrulation would be a good idea. I could re-visit the preliminary finding made by Jonathan Cooke when he filmed gastrulation cell movements, that cells appeared to migrate in pulses. This opened the possibility that cells in higher vertebrate embryos might move by chemotaxis, and even perhaps that they might be attracted by cyclic-AMP. It seemed a good opportunity to link the topic of my dissertation with some experiments on real embryos, and therefore chose this for my thesis work. Brian produced some equipment, probably the same as had been used by Jonathan a few years before: a wartime 16mm Vinten cine camera with a huge, clunky time-lapse controller came out of a cupboard along with a basic inverted microscope and together, we connected them to each other. Brian assigned me the room that had, until just before, been occupied by Lür Willnecker. There was a bacterial shaking incubator in the room which we placed standing upright on the bench – the inverted microscope just fitted inside. Apart from him directing me to a paper by Denis New (New 1955) describing the culture of chick embryos *in vitro*, I received no further experimental instruction from Brian so I proceeded to the library to read about how to explant embryos from the egg and about their early development. I had to teach myself how to culture embryos by reading the papers – there were therefore some unintentional departures from New's original method, some of which I only discovered many years later to be significant improvements. It took many months before I succeeded in growing my first embryos and producing a very short film sequence. It was impossible to see individual cells so I tried to mark regions with vital dyes, but in the end used carbon particles following a paper by Spratt (Spratt 1946). Eventually I was able to visualise movements at the late primitive streak stage, and to obtain an indication that they might indeed be pulsatile, with a period of 2.5 minutes – just like the slime mould. Moreover, when analysing the films it seemed as if the initiation of each pulse of movement spread as a wave from the 'organiser', Hensen's node (the tip of the primitive streak). Brian was very excited by this and helped me with statistical methods to analyse the movies. We wrote a paper reporting the findings, which was published the following year (Stern and Goodwin 1977).

Cyclic-AMP was a good candidate as a chemotactic signal to direct these cell movements, as *Dictyostelium*. Brian, Kees Weijer and I discussed a recent paper (Nanjundiah 1974) in which the author had taken advantage of its chemoattractant properties for slime moulds to explore whether the amphibian embryo might contain a source of cyclic-AMP. An embryo was placed in a field of aggregating *Dictyostelium* amoebae, which were found to migrate to the dorsal lip of the blastopore, the 'organiser' of the amphibian embryo. With Kees, we decided to try this in the chick embryo. Of course we were soon frustrated by the problem that *Dictyostelium* amoebae die at 38°C, whereas chick embryos do not develop at room temperature. Kees and I and a workshop technician played around with different methods to create a very steep temperature gradient but never succeeded. The experiment remained only as a good idea.

Trying to emulate Brian, I now wished to model chemotaxis to test whether a pulsatile signal could account for the patterns of cell movement towards the primitive streak during chick gastrulation. First I had to learn some computing so I took programming courses in BASIC, FORTRAN-IV and ALGOL-68 from the Sussex computing centre. Some of this involved learning to read punched cards and carrying around large boxes of these, containing the programming code and the data. Brian and the neighbouring evolutionary biologists had just purchased a very small Wang computer (with a magnetic cassette tape as storage device) for the fifth floor. This made it possible to write comparable code in BASIC and to make changes directly on screen and this greatly speeded up the initial simulations. These models were fun to construct as well as a great learning tool because they introduced more precision into the biological ideas. They did not, however, produce publishable insights and have therefore remained confined to the inside of my thesis.

Around this time Brian suggested that electrical (ion) currents could play a role in directing cell movements and imparting spatial and temporal information. He suggested that I should travel to Purdue University (Indiana) to work with Lionel Jaffe, who had recently invented a device, called the 'vibrating probe', to measure extracellular ion currents non-invasively. I jumped at the opportunity and was soon on a plane to the USA in the Autumn of 1977. Brian had also asked me to obtain one of the phase-lock amplifiers critical to this device and to learn how to make probes so I could install a similar system at Sussex on my return. With Jaffe, in just a few months, I learned my first principles of electrophysiology and we succeeded in measuring very strong currents emerging from the primitive streak during chick gastrulation; not pulses but a steady current (Jaffe & Stern 1979). At Purdue I also first met Art Winfree, who had an impressive intellect – he showed me the Belousov-Zhabotinsky reaction (a chemical system that makes propagating waves that resemble aggregating *Dictyostelium*) and, together with students from the Jaffe lab and others, we often gathered at his home to watch feature films and make popcorn.

The remaining work for my thesis focused on embryonic regulation (the ability of the embryo to give rise to several embryos when cut into fragments) and the role of Hensen's node in induction, but Brian never showed any interest in any of this. Among the areas explored, I became particularly interested in a paper by Hephzibah Eyal-Giladi, reporting that when an early chick embryo is folded along the future midline a single axis develops along the fold, but when a similar embryo is folded across this axis (so as to appose the head and tail ends), three embryonic axes arise (Eyal-Giladi 1969). The arrangement of the axes was very reminiscent of those found by Vernon French on cockroach limbs. I discussed this extensively with Malcolm Maden, wrote to Vernon French and also discussed the ideas with Laurie Iten during my visit to Purdue. Laurie had been a student with Susan Bryant, Vernon French's collaborator, and had recently joined the Faculty at Purdue – her lab was immediately next to Lionel Jaffe's and we talked a lot. I tried a few experiments which confirmed Eyal-Giladi's findings, but then my time as a postgraduate student was over and I never returned to them mainly because there was no obvious way to find out the cellular or molecular basis for this phenomenon. This remains the case but I hope that at some time in the future someone will rediscover the problem and resolve it.

It was early Summer of 1978 and time to write up and submit my DPhil thesis. I struggled with a clunky mechanical typewriter and many sheets of carbon paper (there was a photocopier in the main library at Sussex but it was awkward to use, very expensive, and the copies were very grey and tended to fade within a few months). Eventually the thesis was typed. I knocked on Brian's door and gave him a copy, bound in a temporary folder. It was almost lunchtime on a Friday. Relieved and looking forward to a few days' rest, I went to the bar for a beer with some friends. I returned to the lab mid-afternoon to collect some things and was surprised to find Brian's copy of the thesis on my bench, with a note on top, in his distinctive, angular handwriting: 'Very nice. Brian'. That was his only comment on my thesis. I wish I had kept the note to show to my own students.

Eventually the time came for my *viva voce* examination. Chris Ford was my internal examiner, and Jonathan Cooke, from the MRC National Institute of Medical Research at Mill Hill, London, the external. The *viva* was to take place in Brian's office (without Brian present). As we entered the room, Jonathan's words were not exactly reassuring: 'Don't worry – you'll get your PhD, eventually'. The examiners asked me to make some major changes including cutting out several sections and I was glad that 'eventually' came only about a month later. I became the first of Brian's postgraduate students to complete the DPhil successfully. By then I had already arranged to join Ruth Bellairs, a highly respected chick embryologist who had pioneered electron microscopy and experimental embryological approaches, as a postdoc at University College London later that year. It was the dawn of Molecular Biology, and I soon discovered that in order to succeed, I first needed to shake off the 'anti-gene' image that would automatically be attached to me as Brian's former student. But I have learned a lot from Brian and because of him, both during my student days and thereafter – he was truly inspirational.

Acknowledgments

I am grateful to Claudio Alonso, Jeremy Brockes, Steve Bunting, Jonathan Cooke, Juan Pablo Couso, Paul Harvey, Conrad Lichtenstein, Andrew Pomiankowski, Andrea Streit, Kees Weijer and Lewis Wolpert for their helpful suggestions and for sharing their own memories.

Brian Goodwin's lab, early Summer 1978. Left to right, standing: Neil Turner, [unknown], Stelios Pateromichelakis, Susan Kirk-Bell, Chris Ford, Brian Goodwin, Audrey [technician/cleaner], Joan [technician/cleaner], Kees Weijer, Christina Smith. Sitting: Lynn Trainor, Claudio Stern, Dave Brill, Marina de Waal Malefijt (then Kees Weijer's girlfriend, working with Deborah and Brian Charlsworth), Wendy Neilsen.

A Long Friendship

F.W. CUMMINGS

Frederick W. Cummings is Emeritus Professor, University of California Riverside, and a lifelong friend and collaborator with Brian Goodwin.

One of my most fortunate friendships was that with Brian Goodwin, a friendship of some thirty-five years, starting at a party in London in 1973. A year or two later, I received a letter from him, asking if I knew of a theoretical physicist who would like to spend a year with the Sussex University Developmental Biology group. Yes, I did. This launched me on a distinctly new intellectual path, inspired by Brian's expertise and enthusiasm. From that time to the present I have been on a very steep learning curve.

That year 1975–76 in Sussex was an active one. The few names of people there at that time with whom I had interesting and enjoyable talks include John Maynard Smith, Gerry Webster, Claudio Stern, Malcolm Maden, and numerous others (whose both names I cannot remember after thirty-five years). Vernon French visited and gave an intriguing talk on cockroach leg regeneration; numerous other colloquium talks that year were very imprinting and exciting.

This visit opened a new career path for me, a new scientific life, for which I will be forever grateful. I have since then pursued a common interest of both of us, involving the coupling of pattern to bio form, and the reverse. One of the many subjects that Brian and I talked a good bit about was: 'Is it sensible, or helpful, to think of an analogy between Einstein's theory of general relativity and the interaction of pattern and geometry in living systems?' The ideas of D'Arcy Thompson regarding shape transformations connecting disparate organisms influenced our thinking.

Brian and I got together quite a few times after 1975, in both England and numerous times in the US. He and I visited back and forth numerous times, England to USA, USA to England. We had most fruitful conversations, sometimes contentious, always in good humour. (By far, most of the fruit just referred to was being bestowed by Brian). This exchange went on until his untimely recent death. He is sorely missed.

An Interview with Brian Goodwin: 1
STEPHAN HARDING

Stephan Harding worked with Brian Goodwin from 1996 onwards on developing the MSc in Holistic Science at Schumacher College, Devon. This interview took place on April 26, 2009, in Brian's garden in Totnes shortly before his death.

Personal history

SH: Why don't we begin with your childhood, and how it was that you became interested in science? Was there something in your childhood that propelled you towards biology?

BG: Well, Stephan, it's going to be quite a journey and it's going to be a great pleasure to talk to you about all this.

I grew up in Canada. I was born in 1931. Therefore I was a child of the depression. Now, I didn't notice that, at least not in any significant way, because when you grow up in a family that actually has enough resources, not wealthy, but just enough resources, you are very protected from the world. And so I grew up in an environment where there were five members of the family, strong family bonds. My mother was a wonderful person from the point of view of holding the family together – strong family bonds, and furthermore, giving us our freedom. So we had this combination of freedom – because she was a Scot, she believed in education, and so education was something quite primary and that meant that it was open sesame – we could do as we pleased – there was no particular direction in which we were being pushed or directed. My father was a mining engineer and so he was into the mining business in Canada, and one of my elder brothers followed along those lines and did mining engineering at Queen's University. But again, there was absolutely no obligation on anybody's part to do that. That's the sort of thing that happens in families, you copy the father or the elder brother, but by the time it came to me – and I was the youngest – there were no particular role models that I had, and so what I remember is being very immersed in nature growing up in Canada, and that was in Eastern Canada, Quebec, and we had plenty of open country around. And of course we lived on the Ottawa River, so that this was a natural playground for us on the river either in winter or summer. In the winter it was frozen, so we could go across on our skis. We could ski through the woods. And then in the summer we had row boats and canoes, and my dad was dead keen on canoes and so he taught us all the tricks of canoeing – in particular looking after your canoe – really looking after your canoe. They were canvas canoes, they were pretty tough, but nevertheless you would never let them touch the shore. You would leap out of the canoe and pick it up. This is what the *coureurs de bois* did, and their canoes were made of birch bark – very fragile. They used to leap out of the canoe before they landed and they would pick them up and carry them off. This is the way the Indian preserved the integrity of the canoe. So we learnt this respect for these ways of moving around through the canoe and other forms of craft, because of course it was the waterways that were the natural highways in Canada. We spent a lot of time on our lakes and rivers and going camping and generally enjoying the whole scene of nature. Now that meant that there was a sense in which I was immersed in nature in a very natural way, but when it came to my intellectual development, what happened was, inevitably, a certain divorce between that intuitive, intrinsic respect and understanding for the forces of nature and a desire to understand life as a kind of principle. So this way of looking at the world in terms of principles and concepts goes very deep in modernity – in modern science. We didn't grow up with a phenomenological, Goethean approach in which you cultivate an intuitive relationship with the world. Now, I did have an intuitive connection with nature, but I didn't cultivate that. What I cultivated then was a scientific approach which, presumably, came partly from the fact that my dad was a scientist – an applied scientist – but more, that this was the way, if you wanted to know about life, then this was the natural direction in which to go. You found out about the details of life. And so I began to study biology. But it's quite interesting that at the beginning my ambition was to become a forester and actually look after the great Canadian forests and to do something practical in that sense, because this was the direction that my dad tended to see things – do something practical – do something that would help. But it didn't take me long to discover that my real path was the one of enquiry into what I like to think of as 'the nature of life'. Now that's a pretty abstract question, but there's something about that whole question of what is the mystery of life that really really intrigued me, and I wanted to find an answer to that.

SH: But what drew you into science? You could have explored that question philosophically or poetically, or artistically. Was it because, as you said, your father was an applied scientist and your family seemed to have a scientific atmosphere?

BG: Well, it's quite an interesting question that, because my mother was very much inclined towards poetry and the arts. She was a musician, she played the piano, and so that was the side of her that was very highly developed. And so that was an example to me of a way of relating to the world through the arts – through music and poetry, through literature and so on. But that didn't strike me as the serious way of doing things. What has gravitas? What has significance? What is going to be respected? And you know, you pick these things up very quickly. And that's why I developed this interest in science and a rather abstract way of formulating the question: 'What is life?'

SH: Did your father not see a contradiction between his mining activities and his love of nature? Did that ever come up in your discussions?

BG: Well, there's a sense in which it was there, because we used to make use of nature for whatever purposes seemed to be appropriate. You know, we would go and cut hemlock boughs to prepare bedding for ourselves – we could make very comfortable beds. You did that every night, and there seemed to be an unlimited amount of hemlock. And the same thing with firewood. You would collect firewood and leave it for the next person, so that you left a mark that you had camped in that site, and you were welcoming other people to go and camp in that site. You tidied it up. You know, you put away the hemlock boughs. You didn't burn them up – that would have made a conflagration – start a forest fire. But you left the campsite neat and tidy. So there were conventions about respecting nature and looking after nature, and my dad had a deep, deep respect for the forces of nature. He was always watchful for a sudden storm if you were on the lake. He would watch for any sudden wind that would be a presage of a change in the weather. So he was constantly aware of these things, but there's no doubt that there was a certain conflict between mining, which is actually stripping the earth of its resources and a deep love of nature, but I don't think he experienced that as a contradiction because nature was so plentiful. It just didn't seem that is was going to run out of resource at that time. So yes, contradiction, but not explicit.

SH: Just going back to before this time, I wanted to ask you if you had any special experiences out in the wild in Canada that you can remember.

BG: Oh yes. I can remember what we would now call mystical or shamanic experiences. But there were of the following kind. We'd be playing in the woods aged eight, nine or ten and just running wild through the woods and loving this freedom. And then I'd come across a big boulder – a big rock – and there was something about this rock that attracted me like a magnet. I remember climbing on top of it and lying down and just having a sense of deep, deep peace and fulfilment. It was a kind of mystical immersion. I wouldn't have described it that way. I just thought: 'That's a special place, I love it'. It has special qualities. And so it had a certain power, and our ability as children to recognise that there are power places and that by entering into a particular relationship with them you were just expanded. So that is probably the most distinctive type of experience that I remember that had this quality of deep relationship to the natural world.

SH: Would you go back to that place over and over again, to re-visit that rock?

BG: Yes, or I would look for another one, because sometimes these places are difficult to find again. I would mark them out in my imagination, saying, yes I must go back there, and if I was in that part of the woods I would do that – I would do precisely that. But it is not as if there was a unique spot – you could find other spots.

SH: So I have this picture of you living on the edge of a vast wilderness, that your house was on the edge of a huge expanse of forest, with lakes and rivers, and that you had easy access to those sorts of places.

BG: It wasn't wilderness, we did not live next to wilderness. We lived next to woodland. But there was quite a bit of farming that went on in Eastern Canada, and so it's not as if you had vast tracts, but that woodland and forest was pretty natural, and it had the qualities, for a child of eight or ten, of being just pristine and being like wilderness. But when I think back to it now I realise that some of the woods had been logged. So it was second growth, because in Eastern Canada there has been quite a bit of interference with nature.

SH: Yes, I see. So the woods had a major impact on you. With respect to your family, in what ways would you say your mother influenced you?

BG: I think there were other complications that came from my mother in relation to the whole notion of patriarchy. By the time I was in my teens I was very aware of the fact that we lived in a patriarchal culture, that my mother was an extremely intelligent woman, and that she was made to suffer.

Now, when I say 'made to suffer' that's a bit of an exaggeration, because it's not as if she was exposed to blatant patriarchy, but she was very sensitised to the assumed superiority of the male, partly because of her background. Her father had been a banker. She grew up in Mandalay. Her father retired early to Aberdeen, so a Scottish family, and so her father was the dominant element in the household. In fact, as far as my mother was concerned it was her mother, my grandmother, who was the one who really embodied wit, intelligence, quickness, lightness, poetry, and understanding. She was streets ahead of her husband. But because of patriarchy, of course, she had to obey his decisions, and this hurt my mother.

Now, my mother never discussed this with me, but parents don't have to discuss these things with their children – you pick them up – and I realise that my mother's reaction to various male visitors who were behaving in particular ways was such that, she made it very clear, that she disapproved. She was a powerful woman, so disapproval was one of these things you paid attention to. So that was another complication, you see. Patriarchy is clearly an element of Darwinism in the sense that progress in civilisation is regarded as dependent upon competition, and competition, particularly by males. *[Phone rings – destroys the thread.]* I saw this culture as dominated by competition. My mother was dead against the war. Competitive interactions amongst males she held as rather despicable. For that to be exalted as a principle of progress in the culture ... Now, she never talked to me about this, she didn't articulate this, but this is my way of interpreting it. So her pain, in terms of patriarchy went into my psyche, and I interpreted this as a really deep criticism of the whole way in which the neo-Darwinist perspective is developed in relation to competition, patriarchy, progress, that whole spectrum. It took many years for me to figure this out because it's quite a complex path to follow and I just had some intuitive keys to this at the beginning.

PART 2

EVOLUTIONARY BIOLOGY AND PHILOSOPHY

Biology without Darwinian Spectacles

BRIAN GOODWIN

B.C. Goodwin, School of Biological Sciences, University of Sussex (1982)
Biologist, *29, 108–112.*

The limitations of the neo-Darwinist approach to problems of biological form are now causing serious difficulties in understanding morphogenesis. The adoption of a more balanced view of biological phenomena could lead to a solution of this centrally important problem, with interesting consequences for our view of biology as a whole.

There are essentially two types of explanation used in science, one in the style of history, the other in the style of logic. The former describes events as chronological sequences, which just happen to occur because of a series of fortuitous or accidental causes. A classic example was the suggestion by Buffon, the eighteenth century biologist, that our planetary system originated from a near collision of the primitive sun with another celestial body that just happened to pass near enough for gravitational attraction to pull out from the sun a great tail of matter which then condensed into separate masses that became the planets. The alternative type of explanation was put forward by Buffon's contemporary, the philosopher Kant, who proposed that the planetary system arose as a result of laws, which govern the intrinsic gravitational properties of rotating masses of gas. This is typical of the rationalist approach to scientific problems, while the other type of explanation is called empiricist. Contemporary physics favours Kant's explanation by necessity (i.e. by law) with important consequences: not only does this make the origin of our planetary system intelligible in terms of general laws; it also means that there are likely to be in the cosmos countless other planetary systems similar to ours where life can originate. Buffon's empiricist description in terms of chance leaves events as simple happenings, with nothing further to understand.

For a number of interesting reasons, contemporary biology is dominated by explanation in the style of history, by empiricism. Thus if you ask the question: why do human beings have five digits in their hands and feet, the answer is: because the common ancestor of the land-dwelling vertebrates happened to have five digits in its limbs, and this pattern has been conserved. But it hasn't been conserved in other vertebrates such as birds (three digits) or antelopes (two digits) or horses (one digit). So it is necessary to add a further historical or contingent factor to the explanation, which is natural selection.

For example, it just so happened that the conditions in which horses evolved (relatively flat grasslands) favoured the development of a single elongated digit in the limbs, which conferred the ability to run fast and escape predators. This style of explanation stems largely from Darwin. Before him, such historical and functionalist types of explanation were considered to be a part of the biological story, as indeed they must be since events that just happen by chance always contribute in some measure to natural processes. But the biological realm, and in particular biological morphology, was considered by many pre-Darwinists to be one that is intelligible in terms of general 'laws of form', not simply or merely in terms of genealogies and family trees describing the accidents of history. Darwin brought into biology an extreme limitation of scientific explanation, which released the subject from certain confusions of idealist (an extreme form of rationalist) thinking in relation to problems of biological form. However, this has now led to serious difficulties in a number of basic questions concerned with the nature of biological organisation, of which form or morphology is an aspect. When faced with such difficulties, it is always a good idea to examine the assumptions, which, while bringing certain objects into focus, make others blurred and indistinct. Biologists wearing Darwinian spectacles can see empirical details very clearly, but they have difficulty in perceiving deeper questions concerned with order and organisation, and so with morphology.

To look at this in more detail, let's go back to the example of vertebrate limb morphology to see the difference between logical and historical explanations. It was recognised before Darwin that there is a basic similarity of vertebrate limb pattern, despite the great diversity of morphology observed in amphibians, reptiles, birds, and mammals. Thus the limb starts with a single bone (humerus or femur), then two bones (radius and ulna or tibia and fibula), then a series of smaller bones (carpals and metacarpals, or tarsals and metatarsals), and finally a series of phalanges in the digits. This general unity of type was expressed by the concept of homology, which recognises a principle of organisation that remains unchanged or invariant in all these limb forms, so that they can be seen or understood as transformations of one another. This idea is like that which is used in physics to understand such phenomena as the motion of bodies under the action of a central gravitational force, for example the movement of planets and comets in relation to a sun. Newton showed that, by using his inverse square law of gravitational attraction, one could deduce that the only forms of motion possible are the conic sections: circle, ellipse, parabola, and hyperbola. A body following one of these forms of motion, say an ellipse, can have its motion transformed into another form, say a hyperbola, if its velocity is increased to a value greater than the 'escape velocity', in which case it goes off into space in a new orbit, never to return (in classical mechanics, at least). Newton's theory made intelligible a variety of forms of motion within a unifying theory. Each pattern is seen as a transformation of the others under changes in what are called initial conditions (e.g. velocity, as described above; or position, the other variable that can be used to transform one form of motion into another). What is achieved by this means is an understanding of how diversity arises within an overall unity. That which is unchanged or invariant in all the different possible patterns of motion is a property, which is common to the conic sections and can be given mathematical expression. This is an example of explanation in the style of logic, and it is what the pre-Darwinian morphologists were seeking by the use of the concepts of homology (similarity of type or form) and transformation, in relation to vertebrate limb morphology.

Darwin was not sympathetic to this type of abstract, logical explanation of biological form, and he suggested that in place of the concept of homology one should use the historical idea of a 'common ancestor' from which related forms are derived by small heritable variations. Those forms, which confer adaptive value on the organism, will be perpetuated, and those which do not will disappear. The whole focus of explanation is thus shifted to the domain of history and particular events, either chance heritable variations in the organism or random changes (so far as organisms are concerned) in the external environment. There is then no logical principle of order or transformation left to grasp, and biology becomes a domain of contingencies, of events that just happen to occur and are perpetuated because they happen to confer on organisms adaptive advantage in the environment in which they happen to find themselves. If physics used this type of reasoning, then cosmic evolution would take something like the following descriptive form: hydrogen happened to be the first element to be formed, presumably because it is the simplest in composition, consisting of one proton and one electron. Helium came next because it is closely related in composition to hydrogen (two protons, two neutrons, two electrons) and furthermore it is very stable. Then a sequence of random condensations of protons, neutrons, and electrons occurred which gave rise to the other elements, but we don't yet know in exactly which order they appeared, such as whether oxygen or nitrogen came first, but we guess nitrogen as it is simpler. In such a description there is no overall structure which defines the logical relationship of the elements, as in the periodic table, and no theory of form and transformation (transmutation) of the elements of the type which is given by quantum mechanics. Such a science would have very little explanatory power. In fact, physics is based upon the logical concepts of form, transformation, and order; and historical descriptions, though interesting and important in relation to particular chronologies, such as the actual sequence of condensation, accretion, or loss of the elements in the evolution of the earth, are secondary to an understanding of logical principles.

The biological analogues of the physical elements are organisms of different form or morphology, and it has been established in this century that the appropriate components for defining the composition of organisms are biological polymers such as proteins, nucleic acids, polysaccharides, etc. The great triumph of genetics and molecular biology in recent years has been to achieve an understanding of how a particular biopolymer, deoxyribonucleic acid (DNA), can generate copies of itself by a self-replication process, and also specify the molecular composition of an organism by copying and translation processes such that particular DNA sequences give rise to particular macromolecules. This explains how organisms inherit the capacities to make specific types of molecule. The biological tradition that focuses on inheritance and particulars as a sufficient basis for the explanation of biological phenomena has given rise to the belief that organisms can be exhaustively described and understood in terms of their molecular composition; that is, in terms of when and what types of molecule are made in the life history of an organism. Since the information for this is considered to be written in the genes, the concept has arisen that organisms are specified by 'genetic programmes' which are the historical records of successful adaptations of the species (ultimately due to specific molecules). A characteristic feature of programmes is that they can contain any instruction whatsoever; that is, there are no constraints at all on the information which may be included in them, and hence no constraints on the molecular composition which programmes can define. And, since the genetic programme is supposed to determine all properties of organisms including their morphology, we conclude that there are no constraints on possible morphologies. This is precisely the freedom allowed within historical explanations, because anything can happen by chance, and natural selection imposes no systematic order on stable or 'successful' morphologies. It is this view of organisms which has now led to serious difficulties in its attempts to explain organisation and morphology, and we must try to see where the approach goes wrong.

Let us ask if it is generally true that composition determines form. In physics and chemistry this is certainly not the case. If I tell you that an atom is made up of one proton and one electron, you can tell me that it is hydrogen but you cannot tell me its form. For this you need to know other things: first, that the possible forms of hydrogen are described by solutions of Schrödinger's wave equation, the wave functions defining the possible electron states of the atom; and secondly, you must know the energy level which specifies a particular solution. Again, if I tell you that I have a crystal made of carbon, you cannot tell me its form because it could be either graphite or diamond. Of course there are instances in which composition *does* determine form, such as common salt which has only one crystal structure and is not polymorphic like crystals of diamond or sulphur. Examples in biology such as viruses in which macromolecular composition determines (largely but not completely) the form of the virus have been extensively studied and strongly stressed as models for the molecular specification of morphology. But there are many examples in biology in which molecular composition does not determine form, because the same molecules (e.g. proteins) can be assembled into structures of different shape. And certainly a knowledge of an organism's genotype is not sufficient to predict its morphology (assuming always a constant external environment). Thus a fruit fly carrying a mutant gene that usually results in a leg appearing where an antenna should be may show the mutant morphology on one side of the body but be normal on the other. If the 'genetic programme' is written in an organism's genes, then this is not sufficient to determine its morphology. Just as in physics and chemistry a knowledge of composition is not in general sufficient to determine form, so is this true in biology. Thus the current theory runs up against serious problems in its attempt to understand biological morphology in terms of history and particulars (i.e. genetic programmes).

The way out of this difficulty is, not surprisingly, the same as that used in physics. In order to describe *form*, which is the ordered distribution of matter and energy in space, one needs a field theory. Schrödinger's wave equation describes the behaviour of electrons in the attractive field of the nucleus, thus defining the space-time form of the electron cloud. It has long been recognised by embryologists that developing embryos have field properties, such as the capacity of a part of an embryo to develop into a whole organism, or the ability of any part of a field such as a bit of the tissue which a limb, to develop into a number of different structures (elbow, wrist, thumb, little finger, etc.) depending upon the particular conditions which act upon it, particularly its spatial relations to other parts. Although these seem to be rather mysterious properties at first sight, there are various ways of understanding aspects of them in terms of such processes as biochemical reactions and diffusion of molecules, electrical fields and currents, or more esoteric processes such as coherent oscillations of electrical charge associated with membranes and polarisable macromolecules (those with moveable electrons). Instead of trying to force organisms into a mould of particulars determined by genes and the molecules they produce, one can discover the field equations which describe their possible forms and then regard particular macromolecules as contributors to the specification of particular field solutions which define morphology.

Since the form of an organism comes about as a result of a systematic series of changes of shape, starting from something relatively simple like a single cell (a fertilised egg) or a seed and ending up as an adult organism of specific shape like a newt or a sunflower, the process of morphogenesis is to be understood as the developing organism taking on or manifesting a particular sequence of field solutions which transform one into the other in time until the adult form reached. Molecular composition influences the appearance of field solutions which define particular morphologies, so that organisms become entities which are defined by both fields and particles (molecules), or more accurately some combination of the two, since they are not separable. The morphological relationships between different organisms can then also be understood in terms of transformations of field solutions, since these all belong to one general category of structure (solutions of field equations). Thus it becomes possible to recover, in mathematical form, the concept of transformation which the pre-Darwinian morphologists were seeking; but this now comes within the context of the generative processes which underly morphogenesis rather than in relation to static adult morphology. Developmental biology, the study of morphological transformations, thus becomes the logical foundation for the understanding of the taxonomic relationships of organisms, that is, their relationships via transformations of field solutions, defining forms. Just as the periodic table of the elements and quantum mechanics define the logical relationships between the elements in terms of form and transformation, so a rational taxonomy of organisms and a theory of their logical or structural relationships should come from a biological field theory. Molecular biology and genetics, which are concerned with the molecular composition and the inheritance of particulars in organisms (capacities to make particular types of molecule), thus provide some of the details about the way in which specific forms are brought into realisation during development by the specification of particular field solutions.

A simple analogy may illustrate this rather abstract type of reasoning which is so unfamiliar to biologists because we have all had such an extremely empirical training, Darwinian spectacles being provided as soon as we start our biological education. When you step out of the bath and pull out the plug, the water runs out with either a right or a left-handed spiral (it doesn't depend on which hemisphere you live in, because it's the water movement you generate near the drain that determines the handedness of the spiral). Suppose you get a right-handed spiral. Then you can stick your hand in the bath, swish the water around with a left-handed motion, and you then get a left-handed spiral down the drain which is every bit as stable as the previous right-handed form.

Something similar happens in biology. There are many spiral forms in nature, such as the arrangement of the little seed-bearing units (scales) in a pine cone or the shell of a snail. If you look at a sample of pine-cones from a single tree, you'll find that the major spiral is right-handed in some, left-handed in others. Both forms, which are generated by the morphogenetic field in the bud that develops into the cone, are expressed with more or less equal frequency, like the spirals in the bath if you keep a tally on a number of cases. The genes of the pine do not have any influence on the handedness of the spiral form produced. However, in the snail *Limnaea*, it is known that the handedness of the spiral shell is affected by a gene, the wild-type or dominant form being right-handed and the recessive left-handed. Evidently, gene products in this organism have the capacity to create a condition favouring one or other spiral form. Biologists, focusing on particulars, tend to conclude that in such cases the gene product generates the form, in accordance with the general proposition that the genetic programme is responsible for producing the observed morphology. Going back to the bathwater, we can ask the question whether or not your hand, transforming a right-handed spiral into a left-handed one, generated the form or was responsible for its production. And now we realise that it is the property of the liquid state as described in an equation called the Navier-Stokes equation that results in the possibility of different field (in this case hydrodynamic) solutions for motion down a drain, left and right-handed. What your hand does is to produce *particular conditions*, which select or stabilise one of these solutions rather than the other. It has certainly not created or generated them. If the bath had been filled with helium or concrete, you certainly would not get spiral flow down the drain. It has to be a liquid whose properties are described by a particular field equation with certain possible solutions, each of which is realised under specific conditions. I suggest that it is the same with the snail or the pine-cone. The particulars, which determine a specific handedness of spiral come from gene products in the case of the snail, and from conditions that just happened to prevail when the spiral was being generated in the case of the pine-cone bud. Just as the properties of a liquid are described by certain field equations, so the properties of the living state are expressed in field equations whose solutions describe the possible forms of developing organisms. These properties are passed on from organism to organism; that is, they are inherited as the general or universal properties of the living state. Genes may or may not *contribute* to the internal selection of particular solutions, but they do not *generate* form. Hence statements of the type that are seen in just about every popular article about genes, such as a recent *Guardian* report on 'The flaws that lead to cancer' which claimed that 'These (chromosome) strands carry the complete set of instructions for all the characteristics of a human being', are simply wrong. Unfortunately, such statements about organisms also occur in most textbooks and in many research articles as well.

There are many consequences of this biology without Darwinian spectacles, not the least being concerned with the way one looks at evolution. Once it is recognised that there are principles of organisation and laws of form in biology, these time independent properties of the living realm become once again central to the subject and details of evolutionary history become secondary, inverting the present emphasis. Morphogenetic fields become the determinants of evolutionary potential, a statement with experimental predictions about the expected variations in organismic morphology against a fixed genotype. That is to say, the realisation that genes do *not* generate biological form leads to a rather different view of the evolutionary process in terms of the potential forms of organisms and their appearance on the earth. At the same time, the whole understanding of organisms as fields of potential (i.e. possibilities) and how these are realised or actualised brings about a dramatic shift of perception concerning the nature of freedom and determinism in the biological realm, since genetic determinism is seen not to operate as commonly described.

In conclusion, I suggest that it is high time biologists escaped from a very cramped and limiting view of organisms and their evolution, which has led to the mistaken idea that organisms are the result of a kind of coded existential history written in their 'genetic programmes'. This extreme empiricist viewpoint allows for the solution of many problems relating to particulars, those myriad differences of form and

53

behaviour between creatures which so attract our attention, but it is unable to describe and articulate in mathematical terms the deeper principles of order and organisation that make the biological realm an intelligible unity. For this we need to take a lesson from mainstream scientific thinking, which teaches us to understand the phenomenal world in terms of both universals and particulars.

It is becoming clearer to an increasing number of biologists (though still an extreme minority) that the way to apply this lesson in their subject is to describe developing organisms in terms of both universal generative fields and macromolecular particulars, which provide the logical foundations for an exact science of biological form and hence an adequate theory about the origin and nature of organisms of specific morphology. This requires a rewriting of the origin of species so that 'origin' is understood primarily in its logical, generative sense, and secondarily in historical terms.

Form and Function in Biology: Placing Brian Goodwin

MICHAEL RUSE

Michael Ruse is a historian and philosophy of science with a keen interest in evolutionary biology. As a committed Darwinian, there was little on which he and Brian Goodwin agreed. If anything, this was the basis of their warm friendship and mutual respect.

One of the best ditties of W.S. Gilbert (of Gilbert and Sullivan fame) comes in the opera *Iolanthe*.

> I often think it's comical
> How Nature always does contrive
> That every boy and every gal
> That's born into the world alive
> Is either a little Liberal
> Or else a little Conservative!

I find it no less comical that every little biologist that's born into the world alive, is either a little formalist or a little adaptive (ist)! And in this insight lies the secret to understanding Brian Goodwin.

Formalism versus functionalism

The division between the formalists and the functionalists starts with the Greeks (Russell 1916; Ruse 2003). Plato was the archetypal formalist. Much influenced by the mathematical theorising of the Pythagoreans, Plato's view of reality stressed always its underlying formal structure. In particular, in his great work the *Timaeus*, Plato showed through and through how he regarded the universe and its contents as informed by and dependent upon formal mathematical structures (Reydams-Schils 2003). For instance, the atomic contents of the universe are all moulded according to the regular solids, like the tetrahedron (four faces), the cube (six faces), and the octahedron (eight faces). Plato explained the painful nature of fire in terms of its being made up of tetrahedra, little particles that are spiky and that thus uncomfortably penetrate the skin.

Plato's great student Aristotle started life as a professional biologist. There is little surprise then that for Aristotle the overwhelmingly distinctive aspect of organisms is the extent to which they are end-directed. Their parts function or work with purpose. The eye, for instance, can only be understood fully in terms of its purpose, namely to see. Aristotle referred to this kind of understanding as 'final cause' understanding. In today's language, Aristotle stressed the 'adaptive' nature of the living world and its parts. He was a 'teleologist'. Aristotle appreciated that there are other kinds of causes, making up what we today would call 'efficient' causes, bringing things about physically. But they complement final causes, they do not substitute for them (Lennox 2001).

Although one should be wary of reading our interests and beliefs too firmly back into the thinking of the Ancients, it is worth noting a difference that emerges, if not so much directly in Plato and Aristotle themselves, then certainly in their followers. Aristotle wanting to see how things work, tended to keep values out of his discussion. The world was not exactly dead for him, anything but, however the world did function in its own right and we have our concerns somewhat separately. For Plato, however, reality is totally value-infused, with a spirit running through and uniting everything. He saw our world as reflecting Ideal structures, what are known as the Forms, and that these Forms are themselves linked into unity ultimately getting their existence from the Form of the Good or the Beautiful. For Plato, the stress is on wholeness, integration, and the essential worth of all of this. You cannot and should not try to get away from what is of value.

Moving the story down to the modern period, we find that functionalism throve above all in the circles of British natural theology. Archdeacon William Paley in his *Natural Theology* (1802) made the end-directed nature of organisms the central plank to his famous retelling of the argument from design for the existence of God. The eye is like a telescope. Telescopes have telescope makers. Therefore, the eye must have an eye maker, the Great Optician in the Sky! Formalism too had its devotees. In Germany, Goethe – and following him a whole school, known as *Naturphilosophie* – made structure central to organic understanding and in Platonic fashion saw underlying patterns that repeat through nature (Richards 2003). Famously, Goethe himself argued that there is a basic plant structure, the *Urpflanze*, from which all other plants are derived. Ultimately there is connection, unity, and value in everything.

One gets the same sort of thinking in the English Romantics – connection, unity, living spirit in everything.

> And I have felt
> A presence that disturbs me with the joy
> Of elevated thoughts; a sense sublime
> Of something far more deeply interfused,
> Whose dwelling is the light of setting suns,
> And the round ocean, and the living air,
> And the blue sky, and in the mind of man,
> A motion and a spirit, that impels
> All thinking things, all objects of all thought,
> And rolls through all things.
>
> (William Wordsworth, *Tintern Abbey*)

The great German philosopher Immanuel Kant (1790) rather tried to straddle the divide between formalists and functionalists. On the one hand, he recognised fully the formal structures that underlie organisms. In particular, he picked out for special attention the shared structures between organisms of quite different types. Shared structures which we today interpret in evolutionary terms as evidence of common descent, and which are known as 'homologies'. On the other hand, Kant recognised equally fully the end-directed nature of the living world. He appreciated that organisms exhibit final causes. Life scientists:

> ... say that nothing in such forms of life is in vain, and they put the maxim on the same footing of validity as the fundamental principle of all natural science, that nothing happens by chance. They are, in fact, quite as unable to free themselves from this teleological principle

as from that of general physical science. For just as the abandonment of the latter would leave them without any experience at all, so the abandonment of the former would leave them with no clue to assist their observation of a type of natural things that have once come to be thought under the conception of physical ends. (p.25)

So impressed was Kant by this need of final-cause understanding, that he inclined to think that it would be impossible to give a full explanation of the living world, since for Kant ultimately everything had to be reduced to efficient causes. Notoriously, Kant said there will never be a Newton of the blade of grass.

After Kant, as we move into the nineteenth century, we find that the formalist-functionalist dichotomy starts to grow and generally speaking, as I quipped at the beginning of this paper, we find that biologists are either essentially formalists or essentially functionalists. For instance, in early nineteenth-century France, we find on the one hand that there is the great comparative anatomist Georges Cuvier who argued strongly that the distinctive aspect of organisms is their end-directed nature (Coleman 1964, p.42). They exhibit what he called 'conditions of existence'. On the other hand, we find the other great comparative anatomist Etienne Geoffroy Saint-Hilaire, who stressed above all the formal nature of organisms, using the shared similarities (homologies) as evidence of shared origins (Appel 1987, p.69).

Coming down to the middle of the nineteenth century and crossing over to England, the formalist-functionalist divide is perfectly exhibited by two great rivals, the anatomist Richard Owen and the naturalist Charles Darwin. Owen made much of the underlying formal similarities between organisms, arguing that they exhibit a shared archetype. In the case of the vertebrate archetype, Owen made explicit reference to its Platonic roots. 'What Plato would have called the "Divine idea" on which the osseous frame of all vertebrate animals ... has been constructed' (Owen 1894, 1, p.388). Charles Darwin made no less of the functional nature of organisms, and it was this aspect of organisms to which his mechanism of natural selection was directed. He believed that he could counter pessimism, showing that unguided laws can speak to final cause (Ruse 2008, p.72).

People on both sides recognised the evidence of their rivals. The tendency however was to downplay the significance of the rival's position. Darwin was entirely typical in this. In the *Origin* he wrote: 'It is generally acknowledged that all organic beings have been formed on two great laws: Unity of Type, and the Conditions of Existence'. However, there is a pecking order.

> For natural selection acts by either now adapting the varying parts of each being to its organic and inorganic conditions of life; or by having adapted them during long-past periods of time: the adaptations being aided in some cases by use and disuse, being slightly affected by the direct action of the external conditions of life, and being in all cases subjected to the several laws of growth. Hence, in fact, the law of the Conditions of Existence is the higher law; as it includes, through the inheritance of former adaptations, that of Unity of Type (Darwin 1859, p.206).

It was in the twentieth century that Darwinian evolution triumphed. Whether or not one thinks this is a good thing, the simple fact is that, with the coming of Mendelian genetics, it was possible to show how natural selection can function as a decisive cause of organic change. To use a hackneyed term, biology finally had its paradigm. And the point to stress is that this paradigm is one that is firmly functionalist. In the language of the late John Maynard Smith (1969), for biologists – especially for evolutionary biologists – the overwhelming problem is that of 'adaptive complexity', and it is to this that natural selection speaks and gives the answer. The eye and the hand came about because those organisms with variations pointing towards the eye and the hand survived and reproduced and those that did not have such variations died without (or with fewer) issue. The winners were naturally selected and eventually there were full-blown adaptations. Organisms exhibit Aristotelian final causes.

Formalists unbowed

However, from the beginning of the twentieth century, there were always those who took the alternative route. They may have been in the minority, but they were articulate and fully biologically informed. Formalism never went underground. It ever had its enthusiasts and representatives. One such person was the late Stephen Jay Gould. Gould is famous (some would say notorious) for his paleontological theory (devised with fellow scientist Niles Eldredge) of punctuated equilibrium. This sees the course of evolutionary history as being one of inaction (stasis) interrupted by occasional sharp and rapid moves to another form (Eldredge & Gould 1972). It is contrasted with Darwinian 'phyletic gradualism' where the course of evolution is seen as regular and smooth, with many intermediates between different forms. Since for Darwin evolution virtually had to be gradual, otherwise one runs the danger of getting out of adaptive focus – random jumps tend to lead to dysfunction – Gould launched a major attack on Darwinian adaptation. In his celebrated paper, 'The Spandrels of San Marco and the Panglossian Paradigm: A Critique of the Adaptationist Programme', Gould (together with fellow Harvard biologist Richard Lewontin) argued against the 'faith in the power of natural selection as an optimising agent', something that 'proceeds by breaking an organism into unitary 'traits' and proposing an adaptive story for each considered separately' (Gould & Lewontin 1979). He didn't want to deny adaptation absolutely, but he queried its ubiquity. Often, he argued, organic features are like the spandrels (triangular areas) found at the tops of columns in medieval churches. They may look as though they have an adaptive function, and perhaps (as in the church in Venice) they are now used for some purpose (in Venice, to offer mosaic portraits of the evangelists), but really they are just non-adaptive by-products of the evolutionary process.

Gould argued that Darwinians go into the process convinced that adaptation is there to be found, and it is therefore little wonder that adaptation is found!

> First, the rejection of one adaptive story usually leads
> to its replacement by another, rather than to a suspicion
> that a different kind of explanation might be required.
> Since the range of adaptive stories is as wide as our minds
> are fertile, new stories can always be postulated. And if a
> story is not immediately available, one can always plead
> temporary ignorance and trust that it will be forthcoming.
> (p.587)

Continuing:

Secondly, the criteria for acceptance of a story are so loose that many pass without proper confirmation. Often, evolutionists use *consistency* with natural selection as the sole criterion and consider their work done when they concoct a plausible story. But plausible stories can always be told.

Gould accused Darwinians of being like Dr Pangloss in Voltaire's *Candide*, always looking on the bright side – 'the best of all possible things, in the best of all possible worlds'. Elsewhere Gould accused Darwinians of spinning any kind of plausible or implausible story. The results are 'just so' stories, akin to the fantabulous stories in Rudyard Kipling's *Jungle Book* – the elephant has a long nose because a crocodile pulled on it, and that sort of thing.

In his own words, Gould wanted:

> ... to reassert a competing notion (long popular in continental Europe) that organisms must be analysed as integrated wholes, with *Baupläne* so constrained by phyletic heritage, pathways of development, and general architecture that the constraints themselves become more interesting and more important in delimiting pathways of change than the selective force that may mediate change when it occurs. (p.581)

Back to Goethe! 'In continental Europe, evolutionists have never been much attracted to the Anglo-American penchant for atomising organisms into parts and trying to explain each as a direct adaptation'. So what is the alternative? It is one that:

> ... acknowledges conventional selection for superficial modifications of the *Bauplan*. It also denies that the adaptationist programme (atomisation plus optimising selection on parts) can do much to explain *Baupläne* and the transitions between them. But it does not therefore resort to a fundamentally unknown process. It holds instead that the basic body plans of organisms are so integrated and so replete with constraints upon adaptation ... that conventional styles of selective arguments can explain little of interest about them. It does not deny that change, when it occurs, may be mediated by natural

selection, but it holds that constraints restrict possible paths and modes of change so strongly that the constraints themselves become much the most interesting aspect of evolution. (p.594)

One person for whom Gould always had great praise was the early-twentieth-century, Scottish morphologist D'Arcy Wentworth Thompson, author of the revealingly titled *On Growth and Form* (1917, second edition 1948). He is deservedly famous today for his insight that different organisms are often founded on the same blueprint, Gouldian *Bauplan*, and then distorted by some mathematical function. But this neo-Goetherian move is only the start. Somewhat oddly praising Aristotle for downplaying final cause, Thompson saw structure as the key element defining organisms.

> To seek not for ends but for antecedents is the way of the physicist, who finds 'causes' in what he has learned to recognise as fundamental properties, or inseparable concomitants, or unchanging laws, of matter and of energy. In Aristotle's parable, the house is there that men may live in it; but it is also there because the builders have laid one stone upon another. (Thompson 1948, p.6)

Sounding very much like a *Naturphilosoph*, he saw the same underlying principles or laws defining organisation or complexity in both the inorganic and the organic.

> Cell and tissue, shell and bone, leaf and flower, are so many portions of matter, and it is in obedience to the laws of physics that their particles have been moved, moulded and conformed ... Their problems of form are in the first instance mathematical problems, their problems of growth are essentially physical problems, and the morphologist is, *ipso facto*, a student of physical science. (p.10)

Thus:

> We want to see how, in some cases at least, the forms of living things, and of the parts of living things, can be explained by physical considerations, and to realise

that in general no organic forms exist save such as are in
conformity with physical and mathematical laws. (p.15)

The shape of the jellyfish gave Thompson a paradigmatic example of
how laws have similar effects across the non-organic-spectrum (see
Figure 1). He saw everything as a straight consequence of the physics
of drops of liquid of one density falling in a liquid of a different,
somewhat lower density. We find exactly the same in the falling liquid
as with the organic, liquid-dwelling jellyfish.

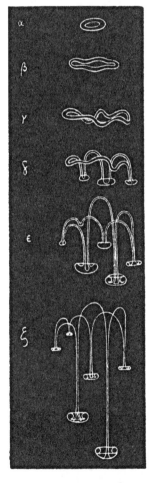

Figure 1. On the left, denser liquid falling through less dense liquid; on the right, a jellyfish. Note the formal similarity, suggesting to D'Arcy Thompson that such intricate forms can be produced simply by the physico-chemical laws of nature without need of natural selection.

> The living medusa has a geometrical symmetry as marked and regular as to suggest a physical or mechanical element in the little creature's growth and construction ... It is hard indeed to say how much or little all these analogies imply. But they indicate, at the very least, how certain simple organic forms might be naturally assumed by one fluid mass within another, when gravity, surface tension and fluid friction play their part, under balanced conditions of temperature, density and chemical composition. (pp.396–98)

Was adaptation involved here? Somehow the question never arises, as it rarely does for more recent enthusiasts for nature's physical laws doing the heavy lifting work. In the words of theoretical biologist Stuart Kauffman:

> The tapestry of life is richer than we have imagined. It is a tapestry with threads of accidental gold, mined quixotically by the random whimsy of quantum events acting on bits of nucleotides and crafted by selection sifting. But the tapestry has an overall design, an architecture, a woven cadence and rhythm that reflects underlying law – principles of self organisation. (Kauffman 1995, p.185)

Form is all important and function is at best on the periphery.

Figure 2. Belousov-Zhabotinsky reaction.

Brian Goodwin

Against this background, Brian Goodwin falls naturally into place. He is the formalist's formalist! Take for instance his fascination with the Belousov-Zhabotinsky reaction, so named after a couple of Moscow-based scientists in the 1950s (Goodwin 2001, p.45). (See Figure 2.) When organic and inorganic liquids are placed on a flat plane (as in a Petri dish) they go through a kind of ring-making exercise that moves out concentrically to the edges. These are very much the kind of movements that one sees in nature itself, particularly the slime molds who behave this way when food gets scarce. Goodwin writes:

> Cells start to signal to one another by means of a chemical that they release. This initiates a process of aggregation: the amoebas begin to move toward a center, defined by a cell that periodically gives off a burst of the chemical that diffuses away from the source and stimulates neighboring cells in two ways: (1) cells receiving the signal themselves release a burst of the same chemical; and (2) the move toward the origin of the signal. (p.46)

As the amoebas go through their movements, leading ultimately to union and then to fruiting and the production of more amoebas, their paths are exactly that found in the Belousov-Zhabotinsky reaction.

The molecules involved in the two cases are quite different, but obviously the feedback systems are parallel. And it is all a question of the way in which nature itself has powers of organisation. Defining a field as 'the behaviour of a dynamic system that is extended in space', Goodwin concludes:

> A new dimension to fields is emerging from the study of chemical systems such as the Belousov-Zhabotinsky reaction and the similarity of its spatial patterns to those of living systems. This is the emphasis on self-organisation, the capacity of these fields to generate patterns spontaneously without any specific instructions telling them what to do, as in a genetic program. These systems produce something out of nothing ... There is no plan, no blueprint, no instructions about the pattern that emerges. What exists in the field is a set of relationships

> among the components of the system such that the dynamically stable state into which it goes naturally – what mathematicians call the generic (typical) state of the field – has spatial and temporal pattern.
>
> (Goodwin 2001, p.51)

It is all form here. Function does not get a look in. Or rather, let me spell things out a little more carefully. Clearly in a sense we have organisms doing what they need to do to survive and reproduce. The point is that what brings this about is not a focus on needs and ends, but rather the unfurling of things according to the internal forces of nature – self-organisation. To use a nice phrase by Stuart Kauffman, we have 'order for free'.

Striking right home at the Darwinians, Goodwin seizes on supposedly paradigmatic examples of natural selection in action, arguing that in fact they are no more (no less either) than the unfurling patterns of nature itself. Take phyllotaxis, the patterns shown by many flowers and fruits in the plant world, for instance the spirals of the sunflower or the twisting lines shown by pine cones (see Figure 3). Darwinians have long argued that these are of direct adaptive significance, usually associated with maximising the amount of sunlight that falls on some specific seed or other. This was the position of Chauncey Wright, early American pragmatist and great supporter of Darwin.

Figure 3. Pine cone showing phyllotaxis.

> To realise simply and purely the property of the most
> thorough distribution, the most complete exposure to light
> and air around the stem, and the most ample elbow–room,
> or space for expansion in the bud, is to realise a property
> that exists separately only in abstraction, like a line without
> breadth. (Wright quoted in Gray 1881, p.125)

A formalist like Goodwin, however, will have no truck with any of this. To misquote Gilbert again, like *'the flowers that bloom in the spring, tra la'*, selection has nothing to do with the case. Goodwin seizes upon the mathematics of the case. The ways of growth force the components into certain familiar grids or lattices, and these in turn are amenable to fairly simply mathematical analysis. The plants in question produce their parts from the center and then push out as they grow. In a sunflower, for instance, one gets one seed and then another and then another – this produces the genetic spiral. As the seeds line up, one by one, new lines or patterns emerge – the most noticeable spirals are known as parastichies. The seeds running along any particular spiral, numbering them in the order they were produced, exhibit fixed patterns. Remarkably the numbers from the crisscrossing parastichies have a formula. The differences between the seeds going the one way (clockwise) and those going the other way (anti-clockwise), follow the sequence: 0, 1, 1, 2, 3, 5, 8, 13... This is the formula – general form, $n_i = n_{i-1} + n_{i-2}$ – worked out by the thirteenth-century Italian mathematician Leonardo Fibonacci, who tried to calculate the number of descendants in any generation from an initial couple of breeding rabbits. The formula is of course better known today as one of the clues in the thriller, *The Da Vinci Code.* (See Figure 4.)

Goodwin argues that that is all there is to it. The developing plants simply follow the rules of mathematics and biological forces have nothing to do with anything. Goodwin is not just Platonic but practically Pythagorean in his numerological enthusiasms. The vulgar fraction series formed by dividing successive members of the Fibonacci series homes in on 0.618, which in turn is what the Ancient Greeks called the Golden Mean, the figure arrived at by dividing the sides of a rectangle such that removing a square from the rectangle leaves one with a smaller but identical rectangle. As it happens, you can get the Golden Mean out of circles also, if you divide up the perimeter properly. This gives you a major angle of 137.5 degrees, which (and if you are not yet convinced

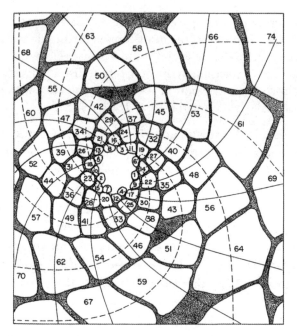

Figure 4. This is a stylized picture of the leaves of a monkey puzzle tree. The leaves are produced, one by one, in the middle and then get pushed out, so the youngest has the lowest number and the oldest the highest number. It can be see that the parastichies contain every eighth leaf produced going clockwise and every thirteenth leaf produced going counter-clockwise. The numbers 8, 13 are successors in the Fibonacci series.

you will be now!) is just the angle on the genetic spiral that divides successive leaves or parts. 'So plants with spiral phyllotaxis tend to locate successive leaves at an angle that divides the circle of the meristem in the proportion of the Golden Mean. Plants seem to know a lot about harmonious properties and architectural principles'. (The meristem is the growing tip of the plant.) (See Figure 5.)

And at this point, following a path for which our history has prepared us, things start to get really interesting. For Goodwin, all of this is but the tip of an iceberg, revealing a philosophy that sees everything as interconnected, in an essentially harmonious fashion, with shared values. As a good Platonist, Goodwin has nothing but contempt for a philosophy that attempts to take meaning and value out of existence. Darwinism is:

> ... an extreme reductionism that makes it impossible for us to understand concepts such as health. Health refers to wholes, the dynamics of whole organisms. We currently

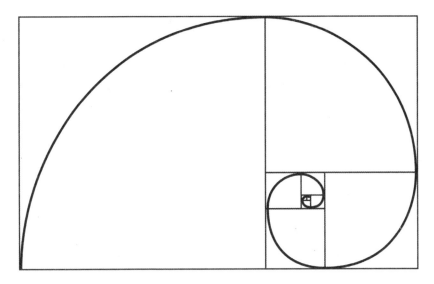

Figure 5. The Fibonacci spiral.

experience crises of health, of the environment, of the community. I think they are all related. They are not caused by biology by any means, but biology contributes to these crises by failing to give us adequate conceptual understanding of life and wholes, of ecosystems, of the biosphere, and it's all because of genetic reductionism.

In an interview with David King in 1996, he says: We have got to escape the Darwinian metaphors of 'competition and conflict and survival' replacing them with metaphors stressing organisms 'as co-operative as they are competitive'. We must turn from 'nature red in tooth and claw, with fierce competition and the survivors coming away with the spoils'. We need a new perspective where 'the whole metaphor of evolution, instead of being one of competition, conflict and survival, becomes one of creativity and transformation' (King 1996).

Expectedly, we have come full circle, returning to Goethe. Talking to John Brockman in 1997, he says: 'I believe that there is a whole scientific methodology that needs to be developed on the basis of what is called the intuitive way of knowing. It's not something that's vaguely subjective and artistic; it's a definite way of knowing the world. In fact, it's absolutely essential to creative science.' Continuing: 'In our own culture, one of the first to develop it was Goethe, towards the end of the eighteenth century.' It is true of course that Goethe's thinking clashed with that of the dominant scientists like Galileo and Newton. But this was to his credit, for it is clear that 'he was developing a different way of understanding the world of phenomena, a way of studying wholes and their relation to parts that can be called a holistic science'. (Brockman 1997)

At least in part this need of Goetherian science seems to stem from Goodwin's conviction that the world is no dead entity, but in a sense throbbing with life, interconnected, working as a whole.

Conclusion

I am a hardline Darwinian. If Goodwin is the formalist's formalist, I am the functionalist's functionalist. I am a mechanist, reductionist, selectionist. I am as Aristotelian as Goodwin was Platonic. But I want to say that even though I disagree right down the line, I see an integrity of thought in the work of Brian Goodwin that I do not see in the work of others with whom I have profound disagreements – in this latter context, I think particularly of the so-called Intelligent Design Theorists and their fellow travellers. Intelligent Design Theory is evangelical Christianity dressed up to look like science in order to get around the separation of Church and State as mandated by the First Amendment to the United States Constitution. It is not just wrong, it is dishonest. That could never be said of Brian Goodwin and his thinking. He really did start from different metaphysical assumptions than someone like me, and everything followed from that. And that is where I am going to leave things here. We have elsewhere both argued our respective cases at length, at very great length. My aim in this short discussion and historical overview has been rather to see why and how two people who care passionately about biology, who are both deeply committed evolutionists, and (I trust it is not inappropriate to say) have burning desires to forward the wellbeing of humankind and of the world within which we live, can nevertheless find themselves so completely in opposition. To achieve true understanding, you do not necessarily have to agree, but you must somehow develop an empathy for the other. I hope my deep affection for Brian Goodwin as a person, my recognition that here was a man of great scientific talent and even greater moral fervour, has indeed let me develop that needed empathy and consequent understanding.

Epigenetics & Generative Dynamics: How Development Directs Evolution

MAE-WAN HO

Mae-Wan Ho was a good friend and colleague of Brian Goodwin for many years, sharing his passion for non-reductionist science and theoretical biology. As Director and co-founder of the Institute of Science in Society and Editor of its quarterly magazine, Science in Society *(www.i-sis.org.uk) since 1999, she has published many books and articles to reclaim science for the public good and to continue promoting the transition to holistic, organic science.*

Abstract and Introduction

More than thirty years ago, Ho & Saunders (1979) proposed the then outrageous idea that the intrinsic dynamics of developmental processes is the source of non-random variations that *directs* evolutionary change in the face of new environmental challenges; and the resulting evolutionary novelties are reinforced in successive generations through epigenetic mechanisms, *independently of natural selection*. The first half of the proposal resonated with structuralism for biology (Webster & Goodwin 1982).

Our proposal has held up well against subsequent research findings, and all the more relevant in view of the numerous molecular mechanisms discovered in epigenetic inheritance (Ho 2009a,b) that could transmit developmental novelties to subsequent generations.

We have demonstrated how the nonlinear dynamics of living processes predicts the major features of macroevolution such as 'punctuated equilibria' (long period of stasis interrupted by abrupt changes); large changes from small critical disturbances, and discontinuous changes from continuously varying parameters; and why macroevolution of form and function is decoupled from the microevolution of gene sequences. We showed that the same (non-random) developmental changes are repeatedly produced by specific environmental stimuli. Furthermore, we demonstrated how general mathematical models can account for all the developmental transformations experimentally produced, which can make strong evolutionary predictions, and offer a natural taxonomy based on the predicted transformations.

However, neither the epigenetic mechanisms nor the dynamics of developmental processes are taken into account in the recent studies on evolution and development.

The totality of research findings gives no support to the neo-Darwinian theory of evolution by the natural selection of random genetic mutations, nor to any theory ascribing putative differences in human attributes predominantly to genes. The overwhelming determinants of health and behaviour are social and environmental. Heredity is distributed over the seamless web of nested organism-environment interrelationships extending from the social and ecological to the genetic and epigenetic. Consequently, there is no separation between development and evolution, and the organism actively participates in shaping its own development as well as the evolutionary future of the entire ecological community of which it is part.

'Epigenetic' then and now

The term 'epigenetic' as used today in epigenetic inheritance refers to effects that do not involve DNA base sequence changes, but only the chemical modifications of DNA or histone proteins in chromatin (complex of DNA and protein that make up chromosomes in the nucleus of cells), which alter gene expression states. Epigenetic inheritance has been defined (Bird 2007, p.398) as 'the structural adaptation of chromosomal regions so as to register signal or perpetuate altered activity states'. But these definitions are rapidly becoming obsolete (Ho 2009a–g). In reality, epigenetic modifications encompass a great variety of mechanisms. They act during and after transcription, and at translation of genetic messages; they can even rewrite genomic DNA (see Ho 2009a). Hence the distinction between genetic and epigenetic is increasingly blurred.

'Epigenetic' as originally used, was derived from *epigenesis*, the theory that organisms are not *preformed* in the germ cells, but comes into being through a process of development in which the environment plays a formative role. Most evolutionists have used 'epigenetic' to mean hereditary influences arising from environmental effects in the course of development.

Evolution: Lamarck versus Darwin

Evolution refers to the natural (as opposed to supernatural) origin and transformation of organisms on earth throughout geological history to the present day. The first comprehensive *general* theory of evolution – that evolution has occurred – was proposed by Jean-Baptiste Lamarck in his book published in 1809, more than two hundred years ago (see Burkhardt 1977; Barthélemy-Madaule 1982; Ho 1983). It was a *uniformitarian* theory in that causes proposed to be operating in the past are the same as those that can be observed at present. The theory postulated the spontaneous generation of the living from the nonliving and unlimited transformation over time, which gave rise to whole kingdoms of organisms beginning from a single origin of life. In addition, Lamarck proposed special mechanisms whereby new species could evolve through changes in how the organism relates to its environment in *pursuing its basic needs*, which produce new characteristics that become inherited after many generations. These special mechanisms are 'use and disuse': *use* enhances and reinforces the development of the organs or tissues while *disuse* results in atrophy; and the 'inheritance of acquired characters', the transmission to subsequent generations of the tendency to develop certain new characteristics that the organism has acquired in its own development.

Thus, Lamarck was also responsible for the first epigenetic theory of evolution, in which development plays a key role in initiating the evolutionary change while specific epigenetic mechanisms transmit the change and reinforce it in subsequent generations (see Ho 1983, 1984a,b).

Darwin's (1859) special theory of evolution by natural selection states that, given the organisms' capability to reproduce more of their numbers than the environment can support, and there are heritable variations, then, within a population, individuals with the more favourable variations would survive to reproduce their kind at the expense of those with less favourable variations. The ensuing competition and 'struggle for life' results in the 'survival of the fittest', so the species will become better adapted to its environment. And if the environment changes in time there will be a gradual but definite 'transmutation' of species. Thus, nature effectively 'selects' the fittest in the same way that artificial selection by plant and animal breeders ensures that the best or the most desirable characters are bred and preserved. In both cases, new varieties are created after some generations.

In addition to natural selection, Darwin invoked the effects of use and disuse, and the inheritance of acquired characters in the transmutation of species. However, those Lamarckian ideas do not fit into the theory of natural selection, and Darwin's followers all regard the lack of a theory of heredity and variation as the weakest link in the argument for natural selection (Ho 1986).

In my book, *Genetic Engineering Dream or Nightmare? The Brave New World of Bad Science and Big Business* (Ho 1997, 1998, 2000, 2007), I have described in detail how Darwin's followers created the 'neo-Darwinian synthesis' by expurgating Darwin's Lamarckian tendencies, including his theory of pangenesis. Darwin's theory of pangenesis had actually received a great deal of support (see Ho 2009h). The rediscovery of Gregor Mendel at the turn of the last century provided evidence that particulate genes controlling the characteristics of organisms are passed on unchanged, except for rare random mutations. This fits in perfectly with August Weismann's (1834–1914) discovery of the material basis of heredity as the 'germplasm' in germ cells that become separate from the rest of the animal's body early in development to ensure it would be protected from environmental influences. Development is therefore irrelevant to evolution. We now know that Weismann's theory is wrong and there are numerous exceptions to Mendelian inheritance. Nevertheless, Darwinism was promptly reinterpreted according to the gene theory in the 'neo-Darwinian synthesis' from the 1930s up to the 1950s and '60s.

As the result of the neo-Darwinian synthesis, evolution occurs strictly by the natural selection of *random* gene mutations, or changes in base sequence of DNA; those that happen to increase reproductive fitness are selected at the expense of the others that do not.

Evolution, development and heredity

The theories of evolution, development and heredity are closely intertwined. Just as evolutionists needed a theory of heredity, so plant breeders in the eighteenth century who inspired Mendel's discovery of genetics were motivated by the question as to whether new species could evolve from existing ones. In accounting for change or transformation, it is also necessary to locate where constancy or stability resides, which constitutes heredity. In order to explain the evolution of form and function, development (epigenesis) is central, as Lamarck clearly grasped. In contrast, Darwin, and neo-Darwinists see new variations arising at *random* in the sense that they bear no direct relationship to the environment; those that happen to be adaptive are selected, while the rest are eliminated. The theory of natural selection is essentially preformist, development playing little or no role in determining evolutionary change (Ho 1984a, b; 1987).

There are a number of different epigenetic theories of evolution since Lamarck; some predating the neo-Darwinian synthesis. A common starting point for all epigenetic theories is the developmental flexibility of all organisms. In particular, it has been observed that artificially induced developmental modifications often resemble (*phenocopy*) those existing naturally in related geographical races or species that appear to be genetically determined. Thus, it seemed reasonable to assume that evolutionary novelties first arose as developmental modifications, which somehow became stably inherited (or not, as the case may be) in subsequent generations.

Epigenetic reorganisation initiates evolutionary change

Early proponents of epigenetic theory included James Mark Baldwin (1896), who suggested that modifications arising in organisms developing in a new environment produce 'organic selection' forces internal to the organism, which stabilise the modification in subsequent generations. Another notable figure was Richard Goldschmidt (1940) who proposed that evolutionary novelties arise through *macromutations* producing 'hopeful monsters' that can initiate new species. In his defence, he pointed out that monsters could be hopeful because of the inherent *organisation* of the biological system that tends to 'make sense' of the mutation. Following Goldschmidt, Søren Løvtrup (1974) advocated a similar theory of macromutations for the origin of major taxonomic groups of organisms such as phyla.

But random mutations – changes in the DNA – that generate hopeful monsters must be hopelessly rare, and to make things worse, major taxonomic groups tend to appear suddenly in clusters, 'adaptive radiations', rather than isolated at different geological times.

The extraordinarily rich fossil finds of the Cambrian 'explosion' responsible for most of the major animal phyla is a prime example of evolution occurring in bursts of 'adaptive radiation' followed by relatively long periods of stasis, or 'punctuated equilibria' (Gould & Eldredge 1972). Furthermore, evolution does seem to proceed top-down, from phyla to subphyla, classes, orders and so on (Valentine 2004), rather than the converse, as predicted by Darwin and neo-Darwinian natural selection of small random mutations. And crucially, all the evidence indicates that macroevolution is decoupled from molecular or microevolution (more below).

These considerations suggest that 'adaptive radiations' involve major novelties arising from *epigenetic reorganisation* provoked by large environmental changes or changes in the organisms' way of life, which also seem to coincide with adaptive radiations. For example, oxygen is very important for the evolution of complex organisms, and the Cambrian 'explosion' is believed to have been triggered by the rapid increase in atmospheric O_2 levels from a low of ~15 % to the current level of ~20 % between 1 billion to 0.5 billion years ago (see Ho 2009i).

'Evo-devo' still blinded by 'genetic programme' of development

In a sense, there is nothing new about the current revival of 'evo-devo' (Gilbert 2003; Carroll 2005; Brakefield 2006; Blumberg 2009; Coyne 2009). It is still dominated by the idea going back at least twenty years that genes control development in a 'genetic programme' of gene regulation and interaction (Coyne 2009); and that large evolutionary changes in body pattern are the result of changes in gene regulation due to natural selection. There is still no recognition that the *patterns* themselves, and biological *form* need to be explained in their own right, independently of whether natural selection operates or not, and independently of the action of specific genes (Ho 1986; Ho &

Saunders 1979; Saunders 1984; Webster & Goodwin 1982). Not surprisingly, there is still little or no recognition that epigenetic and non-genetic environmental influences can give rise to large alterations in form and function.

In a brilliant critique of the genetic determinist approach to behaviour, Gottlieb (1998) deconstructed the idea that genes determine body pattern by pointing to the very different expression patterns of the same *Hox* genes in the fruit fly, the centipede, and the *Onychophora*. *Hox* (homeotic) genes are supposed to control segmental patterning during development; instead, the same genes appear to be simply responding to different patterning processes in the different animals. There is decidedly no homology of genes corresponding to homology of biological structures.

This same theme emerged in a comprehensive review of segmentation in arthropods by Peel, Chipman & Akam (2005), which showed that different groups have distinct modes of segmentation and divergent genetic mechanisms.

It is notable that some researchers now despair of trying to explain pattern formation with complicated computational networks of genes that pass for 'systems biology'. Kondo & Miura (2010) stated that 'the behaviour of such systems often defies immediate or intuitive understanding', and, 'it becomes almost impossible to make a meaningful prediction'.

One important motivation for focussing on development for evolutionary change is that developmental changes are far from random or arbitrary (Ho & Saunders 1979, 1982, 1984; Saunders 1984; Webster & Goodwin 1982); but are determined by dynamical processes, independently of the action of specific genes.

Physicochemical forces and flows in growth and form

Patterns are generated everywhere in the physical world where no genes are involved, and many of the patterns closely resemble those found in the living world. It is the dynamics of physical and chemical forces and flows that generate patterns and forms, as Scottish biologist and mathematician D'Arcy Thompson (1860–1948) so beautifully argued in his classic book, *On Growth and Form* (Thompson 1917). Closer to our time, Alan Turing (1912–1954), English mathematician, logician, code breaker and computer pioneer, is also well-known for his work on morphogenesis (Turing 1952). Turing's reaction-diffusion model shows, for the first time, how patterns can arise *spontaneously* in an initially homogeneous domain, precisely the problem of how patterns can form in a featureless egg in development (Turing 1952; Saunders 1992b, 1993).

The Turing model inspired much work on pattern formation in biological systems before it got lost in the proliferating thicket of genes that 'control pattern formation'. I shall come back to developmental dynamics later.

Waddington's theory of canalisation and genetic assimilation

The most influential figure among the 'epigenetic evolutionists' was Conrad H. Waddington (1905–1975), who attempted to accommodate 'pseudo-Lamarckian' phenomena within neo-Darwinism in his theory of genetic assimilation. Like all Darwinian and neo-Darwinian evolutionists, he wanted to explain the origin of *adaptive* characters, that is, characters that seem to fit the functions they serve.

Waddington (1957) conceptualised the flexibility and plasticity of development, as well as its capacity for regulating against disturbances, in his famous 'epigenetic landscape', a general metaphor for the nonlinear dynamics of the developmental process (Saunders 1990). The developmental paths of tissues and cells are constrained or *canalised* to 'flow' along certain valleys due to the 'pull' or force exerted on the landscape by the various gene products that define the fluid topography of the landscape (see Figure 1). Thus, certain paths along valley floors will branch off from one another to be separated by hills (thresholds) so that different developmental results (alternative attractors) can be reached from the same starting point. However, some branches may rejoin further on, so that different paths will nevertheless lead to the same developmental result. Genetic or environmental disturbances tend to 'push' development from its normal pathway across the threshold to another pathway. Alternatively, other valleys (developmental pathways) or hills (thresholds) may be formed due to changes in the topography of the epigenetic landscape itself.

EPIGENETICS & GENERATIVE DYNAMICS

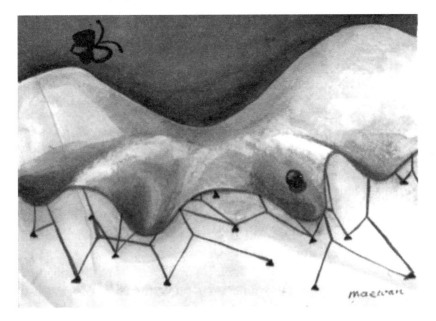

Figure 1. Waddington's epigenetic landscape.

The significance of the conceptual epigenetic landscape is that its topography is determined by *all* of the genes whose actions are inextricably interlinked, and is not immediately dependent on specific alleles of particular genes (Ho & Saunders 1979). This is in accord with what we know about metabolism and the epigenetic system, particularly as revealed by the new genetics (see later). It also effectively decouples the evolution of the organism, of form and function, from alleles of specific genes, and explains the notable lack of correlation between morphological and genetic differences between species (Lowenstein 1986).

Waddington proposed that a new phenotype arises when the environment changes so that development proceeds to a new pathway in the epigenetic landscape, or else a remodelling of the epigenetic landscape itself takes place (both of which are possible from what we now know about epigenetic processes at the molecular level). Thereafter, the new phenotype becomes reinforced or 'canalised' through natural selection for modifier genes so that a more or less uniform phenotype results from a range of environmental stimulus, and later, the phenotype is 'genetically assimilated', so it occurs in the absence of the original environmental stimulus.

Waddington and colleagues carried out experiments showing that artificial selection for the bithorax phenocopy in *Drosophila* induced by ether exposure during early embryogenesis resulted in canalisation and genetic assimilation.

Ho and Saunders' epigenetic theory of evolution

The first distinctive feature of our epigenetic theory of evolution (Ho & Saunders 1979; 1982, 1984) is that neo-Darwinian natural selection plays little or no role, based on evidence suggesting on the one hand that most genetic changes are irrelevant to the evolution of organisms, and on the other, that a relative *lack* of natural selection may be the prerequisite for major evolutionary change.

The second feature is that the intrinsic dynamics of the epigenetic system – developmental dynamics – is determined not so much by gene interactions as by *physical and chemical forces* of nonlinear complex systems in general, which are amenable to mathematical description (Saunders 1984; 1992). That is why, contrary to the neo-Darwinian view, variations of the phenotype that arise during development in response to new environments are *non-random* and *repeatable*.

We proposed, therefore, that the intrinsic dynamical structure of the epigenetic system is the source of non-random variations that *direct* evolutionary change in the face of new environmental challenges. These evolutionary novelties are reinforced (canalised) in subsequent generations through cytoplasmic/epigenetic mechanisms, *independently of natural selection*.

When a population of organisms experience a new environment, *or adopt a new behaviour,* the following sequence of events is envisaged:

a. A novel response arises during development in *a large proportion, if not all of* the organisms in a population experiencing a new environment, due to the intrinsic dynamics of the epigenetic system. In the case of a new behaviour initiated by a single individual in a social group, the behaviour can also spread quite rapidly. For example, Kawai (1962) found that the new habit of washing sweet potatoes in the sea initiated by a young female had spread to the entire troop of wild monkeys on Koshima Island in Japan within nine years. Doubtlessly, this behaviour may also have triggered developmental changes in the monkey's brain.

b. This response is 'canalised' in successive generations through epigenetic mechanisms *independent of natural selection*, and this has been demonstrated in experiments in our laboratory subsequently (see later).

c. After some generations, the response *may* become 'genetically assimilated', in that it arises even in the absence of the stimulus. As in Waddington's epigenetic landscape, this could entail a change in the topography to bias the original branch point in favour of the new pathway, so that the new phenotype will persist in the absence of the environmental stimulus. Random genetic mutations could be also be involved.

Corroborations of Ho and Saunders' epigenetic theory

Since our theory was proposed, we have obtained important empirical and theoretical corroboration. We questioned Waddington's assumption that selection of (modifier) genes is necessary for canalisation and genetic assimilation, and in a series of experiments, Ho *et al.* (1983) demonstrated that canalisation occurred in the *absence* of selection *for* the new character. We showed that successive generations of ether treatment during early embryonic development in *Drosophila* increased the frequency of the bithorax phenocopy in the adults, without selecting *for* the phenocopy. If anything, the phenocopy was almost certainly selected *against*, as it obviously interfered with flight and other normal functions. We had identified a case of 'epigenetic inheritance' of a maladaptive character, consistent with recent findings in 'epigenetic toxicology', in which toxic effects of exposure to environmental pollutants are transmitted to grandchildren (Ho 2009e). At least one study of the fossil record (Palmer 2004) provided evidence that left-right asymmetry in animals and plants may have originated as phenotypic novelties that became genetically assimilated subsequently.

We stipulated that genetic assimilation is not a necessary part of the response to change (Ho & Saunders 1979), as it would preserve the important property of developmental flexibility or 'adaptability'.

In retrospect, this has proved correct. We now know that maternal behaviour, long regarded as genetically inherited and instinctive, is actually associated with epigenetic gene markings that are erased at every generation, yet perpetuated indefinitely from mother to daughter (Ho 2009c).

The complex nonlinear dynamics of the developmental process has been explored mathematically in greater detail (Saunders 1984, 1989, 1990, 1992), and its evolutionary consequences made explicit. For example, it accounts for 'punctuated equilibria' (Eldredge & Gould 1972). It also shows how large organised changes can occur with a relatively small disturbance, and how continuously varying environmental parameters can nevertheless precipitate discontinuous phenotypic change (see Saunders 1990, especially).

The physical and chemical forces that organise living systems were the subject of my book, *The Rainbow and the Worm, the Physics of Organisms* (Ho 1993, 1998, 2008), now in its third enlarged edition. The book presents evidence that cells and organisms are liquid crystalline, with water the most important constituent of the liquid crystalline matrix. I pointed out that electrical polarities determine the alignment of the liquid crystals and hence the major body axes. Furthermore, electro-dynamical forces acting on liquid crystal mesophases may play a key role in pattern formation and morphogenesis. As consistent with this hypothesis, we demonstrated dramatic effects with brief exposures of early *Drosophila* embryos to very weak static magnetic fields; the segmental body pattern of the larva that emerged 24 hours later were transformed into helices (Ho *et al* 1992).

Recently, there has been a revival of interests in electrodynamical processes in development, as changes in membrane potential and the establishment of ionic currents and endogenous electric fields appear to determine polarities long before the relevant genes are expressed (Ho 2011). These and other evidence suggest that electrodynamical processes are involved in pattern formation via the liquid crystalline cortex of the egg and epithelial cells in regeneration.

In contrast, developmental geneticists generally assume that diffusion gradients of special 'morphogens' determine body pattern by providing 'positional information' for particular genes to 'interpret'. For example, in *Drosophila*, where the most complete genetic analysis of development has been carried out (Nusslein-

Volhard 2006), the maternal gene product Bicoid is identified as the morphogen; its antero-posterior gradient serving to initiate the cascade of 'combinatorial regulation' of genes that eventually gives rise to the complete body pattern. The difficulty is that very few molecules diffuse freely in the liquid crystalline matrix, and Bicoid protein is no exception. If anything, it now appears that a gradient of transcription/translation and degradation is actively maintained in the embryo during several cycles of synchronous nuclear divisions (Gregor *et al.* 2007; Gibson 2007), by an as yet unknown patterning process.

Natural selection plays little or no role in Ho-Saunders' epigenetic theory

In our theory, natural selection plays little or no role in evolution (except in the negative sense of deliminating deleterious mutations with large effects) for the following reasons:

1. The epigenetic (developmental and non-genetic) novelties produced in response to new environments are common to most, if not all, individuals in a population, and would swamp out residual effects due to genetic variation.
2. The fluidity of the genome – the constant interaction between genome and environment, the epigenetic markings of genes, and the blurring between genetic and epigenetic – makes clear that organism and environment are inseparable; hence there can be no selection of any static, preformed variant that's independent, or random, with respect to the selective environment.
3. The physical and chemical forces and flows that *generate* biological patterns and forms are independent of natural selection, and require their own explanations (see also Ho 2011).

Neo-Darwinists seem unable to recognise the logical incoherence of applying natural selection to organisms that are changed and changing in non-random ways under the selective regime. Nor do they accept that the generative dynamical forces, which both create and constrain biological patterns and forms, are *independent* of natural selection, relegating natural selection to a negative role of eliminating the unfit.

Instead, they insist that the generative dynamics only provides 'developmental constraints' that limit the action of natural selection to some extent, but natural selection still plays the 'creative' role in evolution (see Bonner 1982).

I shall show why the dynamics that generate patterns and forms are much more than weak 'developmental constraints' to natural selection; and then address the 'neutral mutation hypothesis', the proposal that most, if not all, DNA base changes during evolution are due to random genetic drift decoupled from the evolution of organisms.

Rational taxonomy based on the generative dynamics of biological form

The dynamics of developmental (epigenetic) processes, being amenable to mathematical description, provides a powerful perspective for understanding the development and the evolution of form. That is the basis of 'structuralism in biology' (Webster & Goodwin 1982, Goodwin et al. 1989); or more accurately in our view, 'process structuralism' (Ho & Saunders 1984; Saunders 1984, 1989; 1992, 1993; Ho 1984a, 1988a).

The developmental dynamics define a set of possible transformations that is highly constrained, so that particular transformations may be *predictably* linked to specific environmental stimuli. The fundamental importance of development for evolution is that evolutionary transformations can ultimately be understood in terms of developmental transformations that can be empirically investigated and that this in turn provides us with the criteria for a rational taxonomy, a natural system of classification based on the generative dynamics of form. I shall describe two examples, the segmentation defects in *Drosophila* larva produced by exposing early embryos to ether vapour, and phyllotaxis, the arrangement of leaves around the stem.

The segmentation pattern of the first instar *Drosophila* larva is determined during early embryogenesis. In the course of our studies on the bithorax phenocopy, we discovered that brief exposures to ether vapour also produced characteristic defects in the segmental pattern reflecting a dynamic process arrested at different stages (Ho *et al 1987*).

These defects phenocopy *all* the major genetic mutants identified. And the most general model of successive bifurcation could produce all the observed defects, giving a rational taxonomy of both the observed and yet to be observed forms (Ho 1990, Ho & Saunders 1993). This rational taxonomy based on generative dynamics differs from one based on genealogy or similarity of DNA, and interestingly, also differs significantly from one based on cladistic analysis (Ho 1990). Saunders and Ho (1995) subsequently produced a mathematical model of reliable segmentation based on successive bifurcation.

Figure 2 (Ho 1990, Ho & Saunders 1993) is a transformational 'tree' of the range of segmental patterns obtained *during development*. The main sequence, going up the trunk of the tree, is the normal transformational pathway, which progressively divides up the body into domains, ending up with sixteen body segments of the normal larva. All the rest (with solid outlines) are transformations in which the process of dividing up the body has been arrested at different positions in the body. The patterns with dotted outlines are hypothetical forms, not yet observed, connecting actual transformations.

Figure 2. Transformation tree of body patterns in fruit fly larvae based on a model of successive bifurcations.

This transformational tree reveals how different forms are related to one another; how superficially similar forms are far apart on the tree, while forms that look most different are neighbours. It is the most parsimonious tree relating all the forms.

More importantly, the ontogenetic transformation tree predicts the possible forms that can be obtained in evolution (phylogeny), mostly likely by going up the sequence of successive bifurcations, but occasional reversals to simpler forms could also take place. This is why phylogeny appears to recapitulate ontogeny (Gould 1977). Though actually *it does not*; ontogeny and phylogeny are simply related through the dynamics of the generic processes generating form.

A natural system of classification results from the tree. The twenty-four actual forms or species are classified hierarchically into one 'Family' with two 'Orders', the first Order containing three Genera, and the second Order, eight Genera. The forms not yet found (depicted in dotted lines in Figure 2), would also fit neatly in the natural system of classification should they be discovered in future. There are 676 possible forms according to the dynamic model of successive bifurcation. If all the body segments were free to vary independently, the number of possible forms would have been 2^{16}, or more than 60,000. This demonstrates how highly the generative dynamics can constrain the possible forms, and why, incidentally, parallelisms are rife in evolution (Ho & Saunders 1982, Ho 1984b).

Obviously, the scheme proposed is an over-simplification, which is why 8 hypothetical intermediates (represented in dotted outlines) were not actually observed. The actual process itself may well predict many less possible forms.

In the second example, we produced a transformation tree for all possible ways leaves are arranged around the stem in plants (Figure 3) (Ho & Saunders 1994), based on the generic and robust dynamics that generate the patterns, discovered by French mathematical physicists Douady and Couder (1992). The discovery caused quite a stir in France, as leaf arrangement, or *phyllotaxis,* has been a long-standing problem in biology, ever since Alan Turing drew attention to how the spiral patterns of leaves around the stem conform to the Fibonacci sequence (Saunders 1984 1989, 1992b, 1993). Many neo-Darwinian 'just-so stories' have been invented over the years to account for different leaf arrangements in terms of 'selective advantage'; all of which have been proven irrelevant in one stroke. The power of dynamics – the syntax of form – is that it predicts the set of possible transformations, *excluding all others*. It also tells us how the possible forms are related by transformation (Ho 2008a).

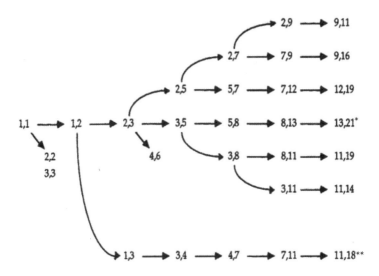

Figure 3. *The transformation tree of possible phyllotaxis patterns.*

It is not known if all the possible forms in Figure 3 actually exist in nature. The main Fibonacci sequence with divergence angle of 137.5° is in the middle row (marked with *). At the bottom is an alternative Fibonacci sequence with divergence angle of 99.5° (marked with **). Like the transformation tree in Figure 2, it makes very definite predictions concerning neighbouring transformations. Thus, parastichies 8,11 and 9,11 (secondary spirals, the numbers indicate spirals to the right and left respectively starting from the centre of the flower or top of the cone), despite their apparent similarity, are quite far apart on the tree, whereas the neighbouring parastichies 8,13 and 13,21 appear superficially very different. As the tree is also an ontogenetic tree, it predicts that plants such as the Canadian pine (*Pinus resinosia*) with parastichies 8,13 in the cone, goes through all of the main sequence in development. We do not know if that is true, but we did find that the leaf shoot bearing the cone has 3,5 parastichies.

For the same reasons, we would predict that the decussate arrangement 2,2 is the earliest divergence from the main Fibonacci sequence, followed by the alternative Fibonacci sequence beginning with 1,3. Phylogenetic transformations are strictly predicted. For example, one would not expect an ancestor of a plant with parastichies 8,13 to have had parastichies 7,12, or even 2,5, but most likely, 5,8.

The dynamics of the processes are subject to contingent 'complexification' (or simplification) in the course of evolution, by virtue of the lived experience of the organisms themselves. Nevertheless, it is highly constrained, when it comes to pattern formation.

It has become clear that directed genetic changes in given environments are just as non-random as morphological changes, and hence, possibly subject to comparable systemic constraints (Ho 1987) (see below).

Natural selection and molecular evolution

Molecular evolution, the study of how proteins and nucleic acid sequences in different species evolve, has been dominated by the neutralist/selectionist controversy that continues to the present day.

Motoo Kimura (1924–94) was best known for his neutral theory of molecular evolution (Kimura 1968), which proposed that most of the amino acid and base changes in evolution resulted from random genetic drift of neutral mutations, i.e., mutations that did not influence the 'fitness' of the organisms. In fact, he did not deny that natural selection could be operating; only that it was not reflected in the evolution of molecules. In effect, molecular evolution appears decoupled from the evolution of organisms, which, at least, is consistent with all other observations indicating the lack of simple translations between genes and phenotype, and is an independent corroboration of Waddington's (1957) concept of the epigenetic landscape.

The neutral mutation theory was inspired by earlier discoveries that when the amino acid or DNA base sequence of genes in different organisms were compared, they diverged apparently linearly according to the time at which the organisms shared a common ancestor. This gave rise to the 'molecular clock' hypothesis (Zuckerkandl & Pauling 1962; Margoliash 1963), according to which, the rate of amino acid or nucleotide substitution is approximately constant per year over evolutionary time and among different species (Lowenstein 1986).

As more data became available, the molecular clock hypothesis ran into trouble. Although there is a correlation between genetic distance and time of divergence, such correlation is not universal, and is often violated.

Numerous studies on extant organisms show that mutation rates are far from constant (Huang 2009). For example, genetic differences between two subpopulations of medaka fish that had diverged for ~4 million years is 3 times that between two primate species, humans and chimpanzees, that are thought to have split 5–7 million years ago. Genetic distances measured on genealogical timescales of less than one million years are often an order of magnitude *larger* than those on geological timescales of more than a million years.

To illustrate the paradox, four randomly selected genes in different species are compared for their similarity (percent identity). All four genes behave as good clocks in macroevolution from fish (*D. rerio*, zebrafish), to frog (*X. laevis*, African clawed toad), to bird (*G. gallus*, red jungle fowl), to mouse (*M. musculus*), and human (*H. sapiens*).

However, they give wildly contradictory timing at lower levels (see Table 1). When different species of fish are compared with each other, *F. rubripes* (puffer fish) vs *D. rerio*, divergence time ranged from 91 to 420 millions of years before the present (myBP).

Protein	Prd	m2 BTK	CytC	GCA1A	Div Time (my BP)
H. sapiens vs D. rerio	39	61	80	66	450
H. sapiens vs X. laevis		65	85	75	360
H. sapiens vs G. gallus	71	85	87	81	310
H. sapiens vs M. musculus	91	98	91	91	91
F. rubripes vs D. rerio	45				420
		71			400
			89		200
				91	91

Table 1. *Genetic distance and estimated divergence time, according to four different proteins. (Huang 2009) Percent Identity.*

Epigenetic complexity vs. genetic diversity, macroevolution vs. microevolution

Huang (2009) proposed that an inverse relationship exists between genetic diversity and epigenetic complexity: multicellular organisms differentiated into tissues and cells are epigenetically complex and can tolerate less genetic variation (germline DNA mutation), whereas single celled organisms, being epigenetically simple, can tolerate more. Consequently, each level of epigenetic complexity will reach its maximum level of variations. This simple theory explains the major features of evolution, including the paradox of an overestimate of divergence times when some gene sequences in lower taxonomic levels are compared (see Table 1).

Humans are undoubtedly the most epigenetically complex species; but in terms of the number of genes, it has only roughly 1.6 times that of a fruit fly and about the same as the mouse or fish. However, the number of certain enzymes responsible for epigenetic gene organisation, such as the PRDM subfamily of histone methyltransferases, increases dramatically during metazoan evolution from 0 in bacteria yeasts and plants, to 2 in worms, 3 in insects, 7 in sea urchins, 15 in fishes, 16 in rodents and 17 in primates. Also, the core histone genes H2A, H2B, H3 and H4 have been duplicated in humans but not chimpanzees, and the number of genes for microRNA (which play key regulatory functions) correlates well with organism complexity. Complex organisms also show complex gene expression patterns: 94 percent of human genes have alternative products or alternative splicing compared to only 10 percent in the nematode *C. elegans*.

For any organism of a certain epigenetic complexity, it can undergo epigenetic changes or genetic mutations in a certain range allowed by the epigenetic complexity. More significantly, epigenetic complexity change is almost by definition, macroevolution, whereas genetic changes due to mutations causing minor variations in phenotypes and do not affect the epigenetic programmes are microevolution. Microevolution, says Huang (2009), is a continuous process of accumulating mutations.

Macroevolution from simple to complex organisms is associated with a punctuational increase in epigenetic complexity and in turn a punctuational loss in genetic diversity. From a common ancestor, the genetic distance between two splitting descendants may gradually increase with time until a maximum is reached and remaining constant thereafter.

The maximum genetic diversity hypothesis predicts that if time is long enough for genetic distance to reach the maximum, then the genetic distance between two genera of the same family should be similar to that between two families, or orders, or phyla. That was demonstrated to be the case for a very old group such as fungi; in contrast, the molecular clock hypothesis predicts that the genetic distance between two fungi genera of the same family should be *smaller* than that between families, and still smaller than that between orders, and so on.

His hypothesis, Huang claims, explains top-down evolution, which is also consistent with the epigenetic origin of evolutionary novelties (see earlier), and the decoupling of macroevolution from the microevolution of genetic distance.

Continuity between epigenetic and genetic changes

Huang's theory does explain a lot and could, in principle, resolve nearly all the major paradoxes in molecular evolution, except perhaps the widely different rates of divergence between different genes within the same organism.

More importantly, I believe Huang's hypothesis that epigenetically complex organisms are less tolerant of genetic or germ line diversity is incomplete, because the level of germ line diversity is *actively* maintained.

A key feature of epigenetics in complex organisms is that they have become more efficient at generating the sequence diversity required at the precise local somatic level (Ho 2009f); and incidentally, also more efficient at reducing it at the germ line level through mechanisms such as gene conversion and concerted evolution (Ho 2004a–d), all part of the death of the Central Dogma of molecular biology, that has been happening since the 1980s.

Epigenetic processes such as RNA editing, alternative and trans-splicing, exonisation and somatic hypermutation, can generate huge sequence diversity wherever and whenever required (Ho 2009f). Some of those processes, coupled with reverse-translation, are powerful mechanisms for generating sequence diversity that can be tested by function within the individual organism, and then used to overwrite the germ line sequence(s). I have reviewed these mechanisms in some detail elsewhere (Ho 2009a), including a range of evidence indicating that mutations are far from random, with the organism choosing when and how to mutate, or not to mutate at all (Ho 2004c).

DNA recoding – rewriting genome DNA – appears to be a central feature of both the immune and nervous systems. DNA recoding is involved at the level of establishing neuronal identity and neuronal connectivity during development, learning and brain regeneration. And it appears that the brain, like the immune system, also changes according to experience.

Mattick and Mehler (2008) suggest that the potential recoding of DNA in nerve cells (and similarly in immune cells) might be primarily a mechanism whereby productive or learned changes induced by RNA editing are *rewritten* back to DNA via RNA-directed DNA repair. This effectively fixes the altered genetic message once a particular neural circuitry and epigenetic state has been established (see Ho 2009f). Steele (2008) has proposed a similar RNA-directed recoding of DNA for the immune system.

Unlike Steele, Mattick and Mehler (2008) fall sort of proposing that the RNA-templated recoding of the genome and the associated structural and functional adaptations could be transmitted to the next generation. This could be crucial for brain evolution in primates leading up to humans, so that the gains made by successive generations could be accumulated (Ho 2009f).

If the analogy with the immune system holds, then as suggested by Steele and colleagues, edited RNA messages or their reverse transcribed DNA counterparts could become inherited via the sperm (Steele 1981; Ho 2009g). 'Sperm-mediated gene transfer' is well-documented as a process whereby new genetic traits are transmitted to the next generation by the uptake of DNA or RNA by spermatozoa and delivered to the oocytes at fertilisation.

Macroevolution therefore involves epigenetic and epigenetically directed genetic changes, and is decoupled from the random microevolutionary accumulation of base sequence changes.

These processes (reviewed in greater detail in Ho 2009a) are part and parcel of the fluid genome (Ho 2003), a molecular 'dance of life' that's necessary for survival (see Ho 2008b, for example).

Heredity and evolution in the light of the new genetics and epigenetics

How should we see heredity in the light of the new genetics and epigenetics? Where does heredity reside if the genome itself is dynamic and fluid? Clearly, heredity does not reside solely in the DNA of the genome. Ten years since the announcement of the Human Genome Sequence brought no progress in understanding life, health or disease. Herculean efforts to locate the genes responsible for common diseases yielded next to nothing (Ho 2010a), not surprisingly at all, given the fluidity of the genome and associated complexity of epigenetic mechanisms.

It has been clear to some of us since before the Human Genome Project was conceived, and copiously corroborated by the findings since: heredity resides in an epigenetic state, a dynamic equilibrium between genetic/epigenetic and other cellular processes. But heredity does not end at the boundary of cells or organisms. As organisms engage their environments in a web of mutual feedback interrelationships, they transform or maintain their environments, which are also passed on to subsequent generations as home ranges and other cultural artefacts (Ho & Saunders 1982, Ho 1984, 1986, Gray 1988). Embedded between organisms and their environment are social habits and traditions, an inseparable part of the entire dynamical complex that give rise to the stability of the developmental process, and which we recognise as heredity (Ho 1984, 1986, 1988b). Heredity is thus distributed over the whole system of organism-environment interrelationships, where changes and adjustments are constantly taking place, propagating through all space-time in the maintenance of the whole, and some of these changes may involve genomic DNA. Thus, the fluidity of the genome is a *necessary* part of the dynamic stability, for genes must also be able to change as appropriate to the system *as a whole* (see Figure 4).

While the epigenetic approach fully reaffirms the fundamental holistic nature of life and discredits any theory ascribing putative group differences in human attributes to genes (Ho 2010b), it also gives no justification to *simplistic mechanistic* ideas on arbitrary effects arising from use and disuse or the inheritance of acquired characters. It does not lead to any kind of determinism, environmental or genetic.

THE NEW GENETICS OF THE FLUID GENOME

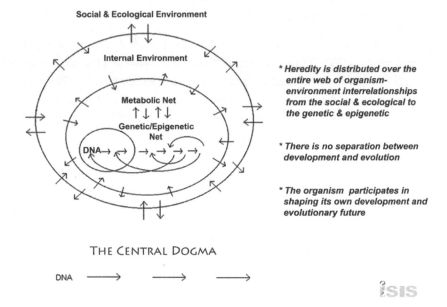

Figure 4. Heredity of the fluid genome versus the Central Dogma.

Organisms are above all, complex, nonlinear dynamical systems (Saunders 1992), and as such, they have regions of stability and instability that enable them to maintain homeostasis, or to adapt to change, or not, as the case may be. The appearance of novelties and of mass extinctions alike in evolutionary history are but two sides of the same coin, we cannot be complacent about the capacity of organisms to adapt to any and all environmental insults that are perpetrated, the most pressing of which is anthropogenic global warming. The dynamics of the developmental process ultimately holds the key to heredity and evolution, in determining the sorts of changes that can occur, in its resilience to certain perturbations and susceptibility to others. And our knowledge in this crucial area is urgently required.

What implications are there for evolution? Just as interaction and selection cannot be separated, nor are variation (or mutation) and selection, for the 'selective' regime may itself cause specific epigenetic variations or 'directed' mutations. The organism experiences its environment in one continuous nested *process*, adjusting and changing, leaving imprints in its epigenetic system, its genome as well as on the environment, all of which are passed on to subsequent generations. Thus, *there is no separation between development and evolution.* In that way, the organism actively participates in shaping its own development as well as the evolution of its ecological and social community. We do hold the future in our hands; it is precious, be careful.

Darwinian Spectacles and other 'Ways of Seeing' Evolution

CRAIG MILLAR AND DAVID LAMBERT

Craig Millar is Senior Lecturer in the School of Biological Sciences, Allan Wilson Centre for Molecular Ecology and Evolution, Auckland, New Zealand.

David Lambert is Professor of Evolutionary Biology and Dean of Research, Science Environment Engineering and Technology Group at Griffith University, Nathan, Australia.

Abstract

As Brian Goodwin remarked in his classic paper 'Biology without Darwinian Spectacles' (see above), explanations in science are essentially of two kinds, either historical or logical. Many branches of sciences use logical principles to order the particular diversity that is the subject of their disciplines. For example, we understand the diversity of elements in the periodic table by reference to a set of organising principles, namely the laws of physics and chemistry. However evolutionary biology is different. Explanations in evolution have been based largely on historical approaches. According to this latter approach, evolutionary biology is made intelligible (explained) through reference to past events and contingencies, rather than any set of underlying laws or principles. Brian Goodwin and others have provided logical as opposed to historical explanations for evolutionary biology. Important players in this rich tradition include Goodwin's PhD supervisor Conrad Waddington and, from earlier times, Johann Wolfgang von Goethe, D'Arcy Wentworth Thompson and Hans Driesch. We summarise these latter approaches and suggest that they are, essentially, manifestations of the same intellectual tradition.

Explanations in art and science

The process of science is often metaphorically regarded as the archaeological uncovering of pre-existing facts. According to this perspective, these facts are buried and simply await discovery. In contrast, artistic creations are regarded as a consequence of spontaneous creativity. Hence the major achievements of art are generally thought to be unique, whilst in science they are regarded as inevitable. Gunther Stent (1978) pointed out that in science, unlike art, discoveries are never seen as contingent. We would say, almost without exception, that what a particular scientist fails to discover today, others will surely achieve tomorrow. Had Darwin not conceived of natural selection someone else would have, and probably at much the same time. Of course Wallace did. It is not surprising then that there exists a relationship between the development of ideas and the context within which they arise. For example, it is no accident that Hegel articulated the dialectical point of view during the industrial revolution – a time of unprecedented social and economic upheaval. Lewes (1856: 156) sums up this view 'discoveries are, properly speaking, made by an age, and not by men' *(sic)*.

Because of the common perception that facts about the world are simply uncovered by the process of science, it is not surprising that our attempts at understanding nature have been based on the gathering of information. Historically, in biology much of this information has been in the form of description. In evolutionary biology such descriptions have been linked by historical rather than logical explanations.

Logical and historical explanations: the death of Captain Cook

The famous explorer and navigator Captain James Cook died at the hands of native Hawai'ian people in 1779 at Kealakekua Bay on Hawai'i (see Figure 1). Although much has been written about this historically important incident, it has typically been depicted as a chance event mediated by some obscure and unfathomable 'reaction' by the 'natives'. The visit to Hawai'i by Cook and the crew of the HMS *Discovery* was regarded by the Hawai'ians as the divine appearance of their lost god-chief Lono who returns annually to renew the fertility

Figure 1. The death of Captain James Cook, 14th February 1779 (Johann Zoffany, fl.1733–1810, Death of Captain Cook, c.1795).

of the land. Lono is associated with natural growth and human reproduction and he returns to the islands with the fertilising rains of winter. But the coming of Lono also represents a threat to the authority of the King. There is a delicate balance between the power of these two important figures, with consequent tensions. Although Cook and his crew were originally greeted warmly by the Hawai'ians, the King was nevertheless surely relieved when Cook left and the *Discovery* set sail in 1779. He had survived the coming of Lono and now the *status quo* could continue as before.

But fateful tragedy struck. The *Discovery* was caught amidst a storm and Cook decided to return to make repairs to the ship's mast. On this return Cook noted that the attitude of the Hawai'ians to him and his crew had changed dramatically for the worse. The inexplicable return of Lono created a 'mythopolitical crisis', as Sahlins put it (Sahlins 1985 p.127). The tensions between Lono and the Hawai'ians were now apparent. It was clear that the former friendship was at an end. Indeed, the delicate relationship was about to undergo an inversion.

The King and the chiefs were anxious to know the reason for Cook's return. They now appeared quite glad to find reason to quarrel with the crew. The King had triumphed when the *Discovery* left, and now the reappearance of Lono acted to interrupt this scene and to challenge his sovereignty. Perhaps Cook and his crew were really intent upon settling on the island and permanently depriving the King of his right to govern? The ritual crisis represented a political threat, and tensions manifested themselves in the form of thefts of items from the *Discovery*.

Cook's policy was not to 'suffer the Indians' else they believe that they had somehow obtained an advantage over him. So after some skirmishes between the crew and the Hawai'ians, Cook decided to take the King hostage, by way of reprimand. Although initially agreeing to come willingly, he suddenly perceived Cook as a mortal enemy. As Sahlins put it: 'this is the structural crisis, when all the social relations begin to change their signs' (Sahlins 1985). Each group reacted in response to their own perceptions of threat and the fatal conflict ensued.

The treatment accorded Cook by the Hawai'ians corresponded to the prescribed sequence of ritual events. 'It was a ritual murder, in the end collectively administered'. Although the killing of Cook was not premeditated by the Hawai'ians, neither was it an accident, 'structurally speaking' (Sahlins 1985). The ritual nature of his death is made clear by the reaction of the priests who, subsequently, went to the ship carrying sections of his hindquarters, and in a state of considerable distress asked 'When the *Orono* (Lono) would come again?' Clearly Cook (Lono) was no ordinary being.

The killing of Cook was finally 'sparked' by a particular event – Cook's dismissal of a old man who was attempting to give him coconut. Cook pushed him and the final conflict ensued. Although this particular account may in some sense be true it does not explain Cook's death. From a European perspective, for example, this seems like an incredibly strong reaction to a rather insignificant act. Why was Cook killed simply for having pushed the old man? Of course, within the context of the whole interaction between two cultures it was an event (a happening of particular significance) in Sahlins' terms. Sahlins argues that the series of events is in an important sense, a consequence of the collision of cultures, and is hence only explained with reference to the dynamic structures (*sensu* the rules and principles by which they operate) of European and Hawai'ian societies.

Yet if the 'coconut man' had not been there, something else would almost certainly have precipitated Cook's death, given that the conflict was at that stage. Indeed a range of contingent events could precipitate the same outcome. Therefore, explanation does not derive from identifying such contingencies. The system was just in a state where a dramatic change was likely to occur. This is why Sahlins remarked that Cook's death was almost inevitable, structurally speaking.

The role of contingencies

Of course, the past history of our world is strewn with contingencies. Had Karl Marx not met Friedrich Engels, had Charles Darwin not happened to read, 'for amusement', the book by the Reverend Malthus on the nature of populations, then the exact history of the world as we know it would almost certainly not be as it is now. The issue is not whether such contingent events have influenced the world around us, but the deeper issue of whether reference to them makes the world intelligible.

Using an historical approach Cook's death is argued to be explained by reference to the above series of contingent events. More generally, within such an explanatory framework, others have attempted to reconstruct the past by connecting events using a narrative (in the same way as in the writing of fiction, see Beer 1983). As Goudge (1961, p.70) pointed out, the aim of the narrative is to make events intelligible by the use of an historical explanation. In fact the narrative is an indispensable tool in reconstructing the past, within this historical approach.

In biology Darwin changed the style of scientific argument by his use of the narrative (Beer 1985). However, the narrative form of explanation cannot be predictive. A number of authors have accepted this point (Goudge 1961; Gould 1986) and consequently argued that prediction should not be the basis of science, but that cause should be. For example, for Gould (1986), 'history causes things'. There is a strong link between narrative, as an art form and history as the causal agent (Beer 1985).

Alternatively we suggest that we can comprehend events (whether biological, social or anthropological) by reference to general principles, in Goodwin's language, by reference to generative processes. Events should be interpreted in the context of these underlying dynamic structures. Complex natural systems generate emergent properties which are not predictable, but which are intelligible with respect to structures. From this point of view understanding is retained even when some contingent events are missing and cause is therefore irrelevant to explanation.

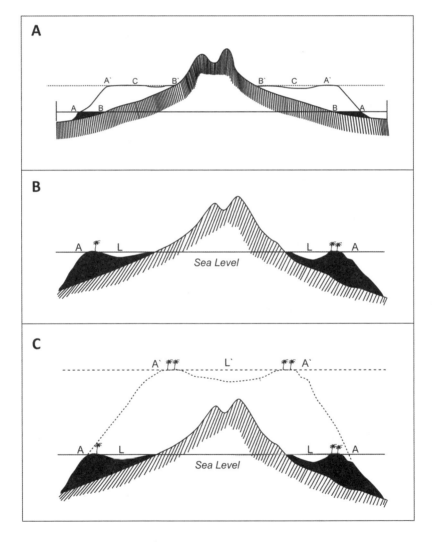

Figure 2. The diversity of coral reefs as a result of the action of geological forces and the growth of corals near the sea surface.

Is evolution best explained by historical narratives or dynamic structures?

Darwin used geological processes to explain the diversity of coral reefs that he observed during his voyage on the *Beagle* (Darwin 1842). The different types of coral reefs were made intelligible, according to Gould (1986), by viewing them as sequential stages of a single historical process. Darwin proposed that volcanoes arise from the seabed to form oceanic islands and later subside. Further he realised that corals only grow down to a particular depth below the sea surface. Consequently, coral reefs begin by fringing oceanic islands, then later become barrier reefs once the island has begun to submerge and finally atolls represent the remains of reefs once the island has completely subsided (see Figure 2). Gould (1986) contended that Darwin's identification of these types of coral reefs into a series exemplifies the idea that you can explain diverse phenomena by identifying an historical order. Hence he suggests that Darwin taught us 'why history matters'.

Of course we now know that Darwin's ordering of coral reef types is correct. Unlike Gould however, we suggest that the sequence is not made intelligible *because* it is an historical order, but because Darwin identified the critical processes of coral growth and their interaction with the geological processes of island formation. The interaction of these processes makes the diversity of coral reefs (fringing, barrier and atoll) intelligible, just as Darwin originally supposed, and not – as Gould has claimed – because history connects them.

Organisms as collections of traits or characters

Darwin's later approach to the evolution of organisms implicitly assumed that organisms could be viewed as being composed of collections of traits. Central to the Darwinian approach is an attempt to explain each of these traits independently of others. In contrast, Goodwin argued against the reductionist atomisation of organisms and his developmental/generative approach was inevitably holistic in nature. Generally therefore, he argued for a logical rather than an historical paradigm in order to approach evolutionary questions. In a very important analysis, Goodwin (1984) questioned the traditional approaches to the long-standing question of the evolution of the

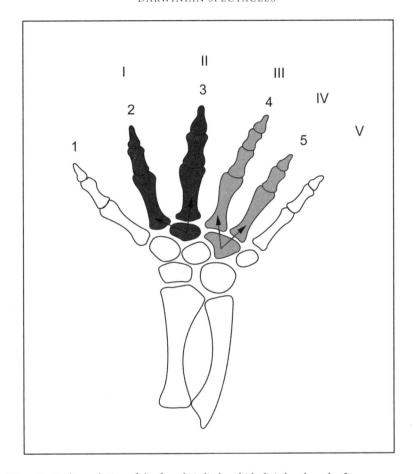

Figure 3. In the evolution of the four digit limb, which digit has been lost?

vertebrate limb. The most common tetrapod limb is the pentadactyl limb, one with five digits. Hence it was assumed that in species such as salamanders with only four digits, there was a gradual reduction in the number of digits, either the anterior (Figure 3, digit I) or the posterior (Figure 3, digit V) one being lost. So the question is which digits remain. However, the question is only appropriate given an atomistic conception of the inheritance of biological form that postulates the inheritance of individual traits (digits) within the pattern observed. Thinking about this problem in terms of generative processes, Goodwin argued that the evolution of one type of limb from another does not involve the loss or gain of individual elements, but the transformation from one field solution to another.

In developmental terms, a field solution with five-fold periodicity is simply a different solution to one with four-fold periodicity. Therefore it does not make sense to label digits in the traditional way, and to ask the question which has been lost? More generally, we can think about the diversity of tetrapod limbs in terms of Conrad Waddington's epigenetic landscape. In doing so we simultaneously conceptualise the diversity of forms in terms of generative processes and resist the temptation to view the organism as a collection of traits.

Goodwin in context

Brian Goodwin's evolutionary views followed in a rich historical tradition beginning with the rational morphologists (Russell 1916). No doubt Goodwin's PhD supervisor Conrad Waddington encouraged his interest in this tradition. His focus on generative processes as explanatory tools to make intelligible biological diversity, was equivalent to attempts by others in the past, notwithstanding his own unique style. For example, Goethe argued that everything was a 'manifestation of something more universal' (Brady 1984). In biology, Goethe's student Geoffroy St Hilaire remarked in equivalent fashion that 'there is intellectually speaking only one animal'. These sentiments are deeply consistent with Goodwin's reference to field solutions. Others have argued similar propositions for example, Lima-de-Faria's chromosome field along the lines of Lancelot Law Whyte. In later times, Gabriel Dover's formulation of the concept of molecular drive has parallels with Susumu Ohno's evolution by gene duplication in the sense that they both utilise the then-known processes that operate at the genome level. D'Arcy Wentworth Thompson's reference to growth fields utilised the underlying mathematical regularities in form and used them to explain evolutionary diversity (see Table 1 opposite).

In conclusion, the central question is, according to the historical mode of explanation does intelligibility in evolution simply 'settle out' of the ordered juxtaposition of events triggered by causal contingent circumstances. We would argue that this view, despite being the dominant view, is problematic. Instead, we would argue that understanding evolutionary biology more properly derives from a knowledge of the underlying dynamic structures in biology.

Acknowledgments

Brian Goodwin was a gentle man, an articulate speaker and a thinker of remarkable ability. We are of the view that his legacy will be considerable and that reading his work will be of great value to generations of biologists in the future.

Table 1.

Some major figures in an intellectual tradition centred on the origin of biological form with diversity understood in terms of transformational principles.

Wolfgang Goethe	The poet-scientist Goethe was a central scientific and literary figure of the nineteenth century. From a biological perspective he argued for the underlying unity of form. He thought that the relationships among diverse biological forms were the result of an underlying dynamic. Therefore the individual entities that comprise this diversity were manifestations of something more universal.
Etienne Geoffroy St Hilaire	Geoffroy was an eminent biologist, a student of Goethe and argued for a 'pure morphology uncontaminated by functional considerations'. He was one half of the 'Great Debate' with George Cuvier in 1830.
William Bateson	William Bateson is famous for coining the term 'genetics' and for drawing attention to Mendel's work. Bateson argued that evolution was fundamentally saltational in pattern and that mutations with large phenotypic effect were responsible.
Hans Driesch	Driesch was an early embryologist and holist. His experimental work was concerned with disturbing embryological development and thereby determining the fate of cells in early embryos. From an evolutionary and philosophical perspective, he formulated a vitalist conception according to which the life processes are governed by an unknowable factor – the entelechy, a term borrowed from Aristotle.

D'Arcy Wentworth Thompson	D'Arcy Thompson's famous book *On Growth and Form* aimed to understand transformations in organismic shape against the background of a Cartesian grid. He noted that in many cases the transformed outlines corresponded closely to the shape of another related animal. Hence he argued that diversity needs to be understood in terms of such developmentally mediated transformations.
Lancelot Law Whyte	Whyte argued that there were important 'internal factors' in evolution and that for a complete explanation of evolution, these needed to be combined with an understanding of external forces that include natural selection.
Conrad Waddington	Waddington was a theoretical biologist and embryologist who attempted to integrate genetics, development and evolution. His visual metaphor of the epigenetic landscape represents an attempt to explain the roles of all the appropriate biological processes.
Susumu Ohno	Ohno postulated that gene duplication plays a major role in evolution in his classic book Evolution by Gene Duplication. He is an active participant in this tradition because he identified a biological processes that shapes genomes. Gene duplication has been shown many times to have shaped the architecture of the genome.
Antonio Lima-de-Faria	Lima-de-Faria has suggested that biological evolution is the last of a number of great evolutions including the evolution of minerals. At the biological level he argued for the importance what he called the 'chromosome field', equivalent to the morphogenetic field of embryology. Again, in the vein of this tradition, he sought to explain diversity at the chromosomal and organismic levels in terms of biological processes.

Brian Goodwin	Brian Goodwin argued for a 'science of qualities'. He had a great interest in archetypes rather than ancestors and aimed to explain biological diversity in terms of generative principles. Brian described his work as following in the classical approach of Paul Weiss, Joseph Needham and C.H. Waddington.
Gerry Webster	Gerry Webster has a strong interest in structuralist biology and the problem of biological form. He has argued for an alternative dialectic between a rational taxonomy of the type suggested by Hans Driesch and a field theory of morphogenesis. Webster argues that current notions of heredity are based on an unexplicated analogy with the social world and that these notions subvert the importance of development and the processes that govern it.
Gabriel Dover	In the 1980s Dover introduced the idea of molecular drive to explain the homogeneity of repeat DNA sequences within species, together with the differences between closely related species. He suggested the molecular processes such as gene conversion and unequal exchange, in combination with sexual reproduction explained this pattern of molecular diversity, rather than natural selection. That is, the biology of repeat sequences was explained in terms of the biological processes that operate at a number of different levels.

From Goethe to Goodwin, via von Foerster

MARGARET BODEN

Margaret A. Boden OBE ScD FBA is Research Professor of Cognitive Science at the University of Sussex. She was a personal friend and colleague of Brian Goodwin since the mid-1960s. Her most recent book is Creativity and Art: Three Roads to Surprise.

Parts of this paper draw on Chapters 2 and 15 of M. A. Boden, Mind as Machine: A History of Cognitive Science *(Oxford University Press, 2006).*

1. Introduction

'Ideas of infinite fruitfulness', worthy of 'immortal renown' ... no biologist could wish for a better recommendation. And coming from the mouth of the greatest scientist of the time? That would be a dream, too good to be true.

In fact, it wasn't a dream. Hermann von Helmholtz (1853), the towering intellect of mid-nineteenth century science, said these things when referring to Johann von Goethe's writing on morphology. And this wasn't unthinking praise, for he was scathing about much of Goethe's work on science – describing his optics as 'absolutely irrational' (Helmholtz 1853: 50).

But maybe it *was* too good to be true. For the hoped-for effect on the world of science didn't follow. Only six years later, Charles Darwin published *On the Origin of Species*, and biologists' interests switched overnight. For both proponents and opponents, the only questions being asked (besides those of mechanistic physiology) concerned evolution by natural selection. Goethe's questions about the origin of biological forms were forgotten – not because they'd been answered, but because they no longer seemed worth asking.

Half a century later, with the discovery of Gregor Mendel's work (and Ronald Fisher's), Goethe's questions appeared to have been answered, at least in outline – but in a way very different from what he'd suggested. The origin of biological forms was now assumed to be a combination of chance, natural selection, and genes. His vision of a 'rational' morphology, although approved by some leading biologists during his lifetime (Merz 1904, ii: 244), could no longer capture attention. In short, neo-Darwinism ruled.

One maverick spirit stood out against this trend, and looked to Goethe for inspiration in so doing. The exceptionally creative biologist D'Arcy Thompson, in his book *On Growth and Form* (1917/1942), insisted that Goethe's questions were still worth asking. He also said that neo-Darwinism was inadequate: not false, but inadequate. His own answers were in some ways broadly similar to Goethe's, but – unlike Goethe's – they boiled down to physics (see Section 3).

He aroused huge admiration, even from orthodox biologists. This was primarily due to his unusual ideas, but also to his style of presenting them: his book was described by one leading biologist as 'beyond comparison the finest work of literature in all the annals of science that have been recorded in the English tongue' (Medawar 1958: 232). In 1942, the eagerly awaited second edition appeared (much expanded, at 1116 pages). It was soon cited by a second towering intellect, Alan Turing (1952), who offered a mathematical theory of morphogenesis that was largely inspired by it (see Section 3).

So had Goethe's ideas triumphed at last? After all, and even though neo-Darwinism was still strong, his example (or anyway, D'Arcy Thompson's) was being followed by a latter-day Helmholtz. Surely, then, Goethe's vision was back on track?

Well ... no. For one thing, Turing's updated version of D'Arcy Thompson's biology could be carried further only by highly competent mathematicians, and only with the help of powerful computers – which didn't yet exist. For another, morphological self-organisation had largely *disappeared* as a scientific problem, surviving only in embryology. The neuroscientist Charles Sherrington had even said that 'were it not for Goethe's poetry, surely it is true to say we should not trouble about his science', and that metamorphosis is 'no part of botany today' (1942: 23, 21).

Even more to the point (and by one of the many ironies of history),

Turing's work on morphology was near-instantly overshadowed, just as Goethe's had been one hundred years earlier. Within a few months of Turing's paper, Francis Crick and James Watson described the structure of DNA; and a few years after that, they deciphered the genetic code. The effect on biologists' interests was even swifter, and even greater, than it had been before. Neo-Darwinism still ruled – but now, largely in the form of molecular biology.

Most molecular biologists saw Goethe's questions as not only not worth asking, but not even intelligible (a change that's fairly common in the history of science: see Jardine 1991). Nevertheless, a few mavericks resisted. The geneticist Conrad Waddington (1940) was one. In the four symposia on theoretical biology that he organised at the Villa Serbelloni in Italy, D'Arcy Thompson's name was often mentioned with admiration (Waddington 1966–72).

However, admiration isn't application. Partly because of the overwhelmingly neo-Darwinist, reductionist, background, but also because of the lack of computer power, the doubts expressed all around the table at Lake Como didn't flower into detailed, testable, theories.

So, 'ideas of infinite fruitfulness'? Maybe, but the fruit hadn't ripened yet.

One of the young biologists invited to sit at the Serbelloni table was Brian Goodwin, a pupil of Waddington's. Goodwin was a skilled experimentalist, with a mathematically literate – and imaginative – mind. His early work, some of which appeared in journals of cybernetics, dealt with circadian rhythms and developmental change (Goodwin 1974, 1976). Over the following years, he put Goethe's and D'Arcy Thompson's ideas back on the biological map again. In so doing, he (with others, such as Stuart Kauffman 1983, 1993) mounted an explicit challenge to neo- Darwinism, and promised to provide a holistic biology that would indeed, if only it could be achieved, earn immortal renown.

This paper tells the story of Goethe's vision of a rational morphology, and asks whether – after all – Helmholtz's evaluation was right.

2. A vision of morphology

It was Goethe who coined the word 'morphology', meaning the study of organised things. It concerns not just their external shape, but also their internal structure and development and, crucially, their structural relations to each other. Goethe intended morphology to cover both living and inorganic nature, even including crystals, landscape, language, and art. But our interest here, like Helmholtz's, is in its application to biology.

In his Essay on the *Metamorphosis of Plants*, Goethe (1790) argued that superficially different parts of a flowering plant – such as sepals, petals, and stamens – are derived by transformations from the basic, or archetypal, form: the leaf. Later, he posited an equivalence (homology) between the arms, front-legs, wings, and fins of different animals. All these, he said, are different transformations of the fore-limb of the basic vertebrate type. Likewise, all skulls are transformations of the archetypal skull. So the intermaxillary bone, which bears the incisors in a rabbit's jaw, exists (in a reduced form) in the human skeleton, as in other vertebrates. (This was significant, because its seeming absence in *Homo sapiens* had been used by others as evidence that God had created a special, unique, plan for the human body.) And all bones, he claimed, are transformations of vertebrae.

In other words, he combined meticulous naturalistic observation (of various vertebrate skulls, for instance) with a commitment to the fundamental unity of nature. Comparative anatomy, on his view, should be an exercise in 'rational morphology': a study of the successive transformations of basic body-plans.

Goethe didn't think of morphological transformations as temporal changes, still less as changes due to Darwinian evolution – which was yet to be defined. Rather, he saw them as abstract, quasi-mathematical, derivations from some neo-Platonic ideal in the mind of God. But these abstractions could be temporally instantiated.

So in discussing the development of plants, for instance, he referred to actual changes happening in time as the plant grows. He suggested that sepals or petals would develop under the influence of different kinds of sap, and that external circumstances could lead to distinct shapes, as of leaves developing in water or in air – a suggestion that D'Arcy Thompson would take very seriously, as we'll see.

The point of interest here is that Goethe focused attention on the restricted range of basic forms ('primal phenomena') in the organic world. He encouraged systematic comparison of them, and of the transformations they could support.

He also suggested that only certain forms are possible: we can imagine other living things, but not just *any* life-forms. In a letter of 1787, he wrote:

> With such a model [of the archetypal plant (*Urfplanze*) and its transformations] ... one will be able to contrive an infinite variety of plants. They will be *strictly logical plants* – in other words, *even though they may not actually exist, they could exist*. They will not be mere picturesque and imaginative projects. They will be imbued with inner truth and necessity. And the same will be applicable to all that lives. (Quoted in Nisbet 1972: 45; italics added).

Similarly, in his essay on plant metamorphosis (1790), he said: 'Hypothesis: All is leaf. This simplicity makes possible the greatest diversity'.

Critics soon pointed out that he overdid the simplicity. He ignored the roots of plants, for instance. His excuse was telling:

> It [the root] did not really concern me, for what have I to do with a formation which, while it can certainly take on such shapes as fibres, strands, bulbs and tubers, remains confined within these limits to a dull variation, in which endless varieties come to light, *but without any intensification [of archetypal form]; and it is this alone which ... could attract me, hold my attention, and carry me forward.* (Quoted in Nisbet 1972: 65; italics added)

Questions about such abstract matters as the archetypal plant were very unlike those being asked by most physiologists at the time. If a body is not just a flesh-and-blood mechanism, but a transformation of an ideal type, how it happens to work – what Helmholtz called its machinery of cords and pulleys (see Section 3) – is of less interest than its homology.

Indeed, for the holist Goethe the mechanism may even depend on the homology. Perhaps it's true that a certain kind of sap – that is, a certain chemical mechanism – will induce a primordial plant-part to develop into a sepal rather than a petal. But what's more interesting, on this view, is that sepals and petals are the structural possibilities on offer. How one describes the plant or body part in the first place will be affected by the type, and the transformations, supposedly expressed by it.

It's not surprising, then, that Goethe was out of sympathy with the analytic, decompositional methods of empiricist experimentalism. By the same token, anyone following in his footsteps – as both D'Arcy Thompson and Goodwin eventually did – would be swimming against that scientific tide. To keep their heads above water, avoiding being the rollers of orthodoxy breaking around them, they'd need experiments too – and, not least, a better understanding of physics.

3. Physics comes on the scene

Helmholtz, although he appreciated Goethe's morphology, really wanted it to be boiled down to physics. As the leading materialist of the time, he was uneasy with any so-called explanation that implied, or even left room for, non-physical causes. Goethe, to be sure, hadn't explicitly posited non-physical causes. But his cryptic remarks on morphological transformations could easily be read in that way.

Helmholtz's ringing endorsement of Goethe's biological ideas was significant not least because he was scathing about much of Goethe's work on science. He scorned his writings on colour vision, for instance, as 'absolutely irrational' (Helmholtz 1853: 50). Instead of 'the poet's' view of nature as 'but the sensible expression of the spiritual', he urged the scientist's view, which was 'to try to discover the levers, the cords, and the pulleys which work behind the scenes, and shift them'. Something was lost in the exchange, he admitted:

> Of course the sight of the machinery spoils the beautiful show, and therefore the poet would gladly talk it out of existence, and ignoring cords and pulleys as the chimeras of a pedant's brain, he would have us believe that the scenes shift themselves, or are governed by the idea of the drama. (Helmholtz 1853: 50)

Helmholtz' point wasn't that *only* lever-and-pulleys explanations are allowed, for he himself posited 'unconscious inferences' in perception (Boden 2006: 6.ii.e). But he took it for granted that those inferences are somehow grounded in 'machinery', alias neurophysiology. Similarly, if he praised Goethe's insights about biological form, he tacitly assumed that a scientific physiology would be needed to understand what was really going on. His caustic remark about 'the scenes shift[ing] themselves' was a criticism of the Romantic philosophers' frequent references to quasi-magical self-organisation in nature. Self-organisation there may be (at least in biology), but to understand it we shall need physiology – and, ultimately, physics.

The first person to take up this point in a serious manner was D'Arcy Thompson. As remarked above, his work was largely inspired by Goethe's rational morphology. He, too, spoke of (abstract/mathematical) transformations of across species of shared forms, such as skulls and body-shapes. But he added a number of intriguing suggestions about the ways in which basic physics (and physical chemistry) may constrain the development of bodily form. For example, he said, surface tension is a significant constraint on the development, and the adult life, of a water boatman – an insect so light that it can literally walk on water. By contrast, when considering the shape of an elephant we can ignore it: gravity is what's important, here. Surface tension might have significance for some of the elephant's internal organs: the alveoli of the lungs, for instance. But it's irrelevant in explaining the creature's overall bodily form.

It's worth noting, here, that D'Arcy Thompson was considering very *general* aspects of physics. To some extent, that was inevitable. Very little was known, in the first decade of the twentieth century, about the specifics of physical chemistry – and virtually nothing about what's now called biochemistry.

That remained true for many years. It was still true in the late 1940s, when Turing started to focus on biology. In the last paper published in his lifetime, he based a new theory of morphogenesis on the basic phenomena of chemical diffusion and chemical reaction (Turing 1952). *Which* chemicals were diffusing in the egg or embryo, he didn't know, nor *how fast* they were doing so. Nor, of course, did he know *what* reactions the chemicals would engage in when they met. But the key point was that, under certain (unknown) circumstances of diffusion-speed and concentration, they would, inevitably, meet. And *if* two chemicals interacted so as to build each other up, or break each other down, then certain patterns of differential concentration would result. Assuming that the chemicals concerned were able to prompt changes in form (what Turing called 'morphogens'), this purely mechanistic self-organisation could lead to visible changes of colour and/or shape and/or colour, such as spots and stripes; petals, tentacles, and segments; and even gastrulation. In short, Turing showed that morphogenesis is (in principle) predictable in terms of reaction-diffusion equations – based not on new biological experiments, but on what was already known about chemical diffusion *in general*.

Turing's paper intrigued and excited many people at the time. Embryologists were intrigued, and so were cyberneticians – including Heinz von Foerster. For the embryologists glimpsed the possibility of making their vague talk of (unknown) chemical 'organisers' more precise and systematic; and the general *(sic)* nature of dynamical self-organisation was a key problem for cybernetics (Von Foerster 1950–55).

As remarked above in Section 1, however, Turing's new theory *did not* turn most biologists towards physics, nor even towards issues of development and morphology. Their attention was diverted, instead, by the discovery of the double helix in 1953, and of the three-letter genetic code soon after that. To be fair, even Turing himself couldn't have done much more, because, as he recognised, he didn't have the computer power that would be needed to study the development of pattern from *pre-existing pattern* (as happens in the successive stages of embryogenesis) as opposed to *pattern-from-homogeneity* (as happens, near enough, in the fertilised ovum). Over the next half-century, then, physics was sidelined and molecular biology reigned.

4. Goodwin and structuralist biology

Today's 'structuralist' biology is highly unorthodox (Goodwin & Saunders 1989; Kauffman 1993). Partly, that's because it employs mathematical complexity theory to develop its theories (Solé & Goodwin 2000), and advanced computer technology to explore their implications. Partly, it's because it is holistic rather than reductionist in approach. And partly, most unorthodox of all, it's not neo-Darwinist.

Structuralists see Darwinian natural selection as a secondary factor in the explanation of biological form. As D'Arcy Thompson put it, what natural selection does is merely to prune the forms that have arisen as a result of physics and chemistry. On this view, physics determines not only what forms are even *possible* (a point which all biologists would have to allow), but also which are most *likely*, so turn up over and over again, across the phylogenetic tree.

So the structuralists reject neo-Darwinism's emphasis on accident and history (Goodwin & Saunders 1989). They *don't* see genetic regulation as an exquisitely finely-balanced process of control that can be sent disastrously off course by the tiniest accidental alteration in the instructions. The awesome complexity of multicellular morphogenesis is admitted – indeed, celebrated. But part of what it means to say that actual biological forms are more 'likely' than (most of) the non-existent ones is that, once they've arisen, they are relatively stable (Goodwin *et al.* 1993).

In other words, some biological forms are 'attractors', which draw dynamical processes into them from many differently-detailed pathways. And this avoids – indeed, denies – the emphasis on accident and contingency that is central to neo-Darwinism. As Kauffman puts it:

> [We] suggest that ... morphogenesis may be deeply robust. Organisms, rather than being tinkered-together contraptions, may exhibit a nearly inevitable and stable order. One's naive intuition might be that reliable occurrence of an ordered morphology from a richly integrated developmental mechanism would require exquisite control of all the variables and parameters of the subsystems making up the integrated system ... *High genetic precision would appear to be required to choose reliably among these many [possible] forms. [Our] suggestion is that intuition is quite wrong.* (Kauffman 1993: 637, italics added)

Goodwin's work confirms this view. Consider, for instance, his mathematical model of the development of the unicellular alga *Acetabularia* (Goodwin & Trainor 1985; Brière & Goodwin 1988; Goodwin & Brière 1992; Goodwin 1994: 88–103). Using computer graphics, this presents the numerical results of mathematical calculations as diagrams/pictures of the developing forms concerned. Like structuralist models in general, it illustrates metabolic/developmental functions, and how they change over time.

Specifically, it simulates the cell's control of the concentration of calcium ions in the cytoplasm, and how this affects, and is affected by, other conditions – such as the mechanical properties of the cytoplasm – so as to generate one type of morphology or another (e.g. stalks, flattened tips, and whorls). It contains some thirty or more parameters, based on a wide range of experimental work. They reflect factors such as the diffusion constant for calcium, the affinity between calcium and certain proteins, and the mechanical resistance of the elements of the cytoskeleton. The model simulates complex, iterative, feedback loops wherein these parameters can change from moment to moment.

One welcome result of running this model was the appearance of an alternating pattern of high/low calcium concentrations at the tip of the stalk, interpreted by Goodwin as the emerging symmetry of a whorl. This result was welcome largely because whorls aren't found only in *Acetabularia*. To the contrary, they are 'generic forms': they are found in all members of this group of algae, and in many other organisms, too. In Goodwin's model, it turned out that whorl symmetries were very easy to find. They didn't depend on a particular combination of specific parameter-values, but emerged if the parameters were set anywhere within a large range of values.

However, the nature of the computer graphics, whose diagrams were composed of many tiny lines – prevented the emergence of visually recognisable whorls (Goodwin 1994: 94). A whorl is a crown of little growing tips, each of which then develops into a growing lateral branch. To simulate this, each of the little tips (calcium peaks) would have to behave (to soften, bulge, and grow) like the main tip, but on a smaller scale. In principle, that's possible. But it would need the machine to draw even more, and even tinier, lines. So it would be very computationally costly, requiring the whole program to be repeated on a finer scale (to generate each lateral branch), and many times over (to generate many laterals).

This failing of Goodwin's simulation was due to lack of computer power. Others were due to lack of biological knowledge. For instance, in each 'run' of the model, the values were static. But in real algae, the parameters can change during morphogenesis, as a result of changes in enzymes. These alterations in the dynamics weren't modelled, because no-one knew just which genes are involved or just how they affect the physics and chemistry of the developing cell. A second flaw was that the model failed to generate the umbrella-like cap of *Acetabularia*. Instead, it gave rise to a bulbous terminal structure. Goodwin's explanation for this was that umbrella caps are not highly probable generic structures: they are less easy to find, within the overall parameter-space of *Acetabularia*. Indeed, they are biologically much rarer than whorls, and they evolved much later within the order of algae concerned, the Dasclyadales (Goodwin 1994: 96f.).

Goodwin had hoped not only to understand the emergence of structure in *Acetabularia,* but also to show that some biological forms are more likely than others. The model's success in finding whorls and failure to find umbrella-caps confirmed his view that he – and, more to the point, biological evolution (including natural selection) – wouldn't have to experiment with a near-infinite number of numerical parameters to find the one-and-only set which would generate whorls. In effect, whorls would be produced 'for free' because they are intrinsically likely.

In his experimental simulations, that is indeed what happened. Proving mathematically that it must happen, that various generic forms must arise in biology over and over again was another matter. But this was/is the ultimate goal of structuralist biology. In Goodwin's words, future work would aim 'to generate by computer simulation the range of forms represented within the order of unicellular green algae using an extended version of [this model], and to explore the sizes of the different basins of attraction of the various genera, and possibly even species, that can arise' (1994: 102). And he added: 'Some of these might turn out to be possible forms that are not represented in either fossil or living species', in other words, what A-Lifers would call 'life as it could be' (see Section 3).

If that could be achieved, the result would be a 'rational' taxonomy, a theory of biological forms 'whose equivalent in physics is the periodic table of elements, [which is based on] the dynamically stable relations between electrons, protons, and neutrons' (1994: 103). Taxonomic similarity would be measured in terms of the parameter space specified by the mathematical theories, and the computer models, concerned. Biology would thus have 'a theory of organisms as dynamically robust entities that are natural kinds, not simply [as implied by neo-Darwinism] historical accidents that survived for a period of time' *(ibid.).*

This biological heresy (to which more and more converts are flocking) provides no comfort for creationists. It concerns the *importance* of natural selection, relative to other scientific explanations, not its truth. D'Arcy Thompson himself didn't deny natural selection (still less, evolution), and neither do his modern successors. Nor do they deny the crucial influence of genes, although they offer a less reductionist, less deterministic, interpretation of gene-action. And nor do they appeal to mysterious non-physical forces to 'explain' biological form. Even Goodwin, whose philosophy of biology and human life goes way beyond most of his colleagues (Goodwin 2006), grounds his eccentric position in strictly scientific experiment and theory. Moreover, the theoretical coherence of structuralism is proven by mathematics and computer modelling.

5. Conclusion

Despite being a committed experimentalist and a rigorously mathematical scientist, someone whom one might have expected to have little sympathy with Romantic *Naturphilosophie*, Goodwin took Goethe seriously. He even co-authored a book of which no fewer than 130 pages were devoted to idealist philosophy (Webster & Goodwin 1996). So, with respect to biology no less than to philosophy and literature, Goethe lives.

But if Goethe's renown is immortal, as Helmholtz predicted, he himself wasn't. And neither was Goodwin. To my personal grief – he was a longtime friend – he died (in July 2009) shortly before this paper was written.

Goodwin left us with many intriguing but unanswered questions, and many more not even asked. It's not unreasonable to say, after Helmholtz, that his biology contains ideas of infinite fruitfulness. Indeed, a leading philosopher of biology remarked that he had 'done the impossible', enabling him 'to understand what idealists are doing' – and added: 'If ever there is a time that the intimations of a science of organic form sensed by certain biologists from Aristotle and Owen to the present are to be realised, it is now' (Hull 1998: 587, 595). In any event, this paper is dedicated to his memory.

An Interview with Brian Goodwin: 2

STEPHAN HARDING

Brian on Maths and Philosophy

SH: To do developmental biology you had to be a good mathematician presumably?

BG: No, not in those days.

SH: I want to back track here. I want to know when it was you first became excited by mathematics, because after all mathematics became a very important part of your work.

BG: Mathematics came very easily to me. It wasn't an obstacle, but it wasn't something I was enchanted by. I'll tell you why I cultivated mathematics. It's directly connected with this rejection of Darwinism. It goes like this: I felt that having been given this basic principle of biology, that this is the basic principle of creativity in evolution. You know, you get spontaneous expression of variation and then competition and then survival of the fitter races is what produces these different varieties. Well, I thought to myself that this is a pretty shallow kind of science compared with what I experienced in physics, because of course in Canada it's a Scottish kind of education and you do a broad spectrum of subjects. I did maths, physics, chemistry at school plus biology and English literature and geography, and all the rest of it – a broad spectrum. So my feeling was that there were principles in physics that said: 'This is the way the periodic table is organised – look at this.' I thought to myself : Bloody hell, that's absolutely fantastic. Here you have some principles that say why the elements are organised in this particular way. Intuitively I thought to myself: well, there's something similar going on with respect to taxonomy, with respect to evolution and the classification of different forms. I thought to myself that I want to understand the principles that could underlie evolutionary biology as a creative expression of coherent wholeness. For that I felt drawn to mathematics – physics and mathematics, because I thought that biology is a subject without principle – it's got no principles. It's funny that, as a way of formulating it.

SH: I see. You formulated that when you were an undergraduate?

BG: Yes, yes.

SH: But it probably wasn't normal for biologists at that time, or even now, to be interested in Whitehead's philosophy, which is a form of panpsychism, isn't it?

BG: Yes, when you take it to the extreme.

SH: So there's a matter aspect and a mind aspect and they are separate, but they are both present in each entity?

BG: They are both present. In a sense they are inseparable – you cannot have one without the other.

SH: But they are distinct in some way?

BG: But they are distinct. He (Whitehead) put the two together in a way that is very ingenious. Well, you know the way that Christian de Quincey does it. He talks about past matter, present mind. In other words every moment of being is one in which there is an expression of creativity , and that creativity exhausts itself in a form which leaves its residue as matter. The creativity is manifest by means of this intelligence, this imagination. So you go from the expression of creative freedom to the expression of something that is exhausted in matter. This is the way he describes it. It leaves its traces in matter, which acts as a constraint for the next moment of creation. This is the implicate and explicate order.

SH: So matter is a kind of secretion of creativity?

BG: That's right! Yes, yes.

SH: Like a crab making its shell.

BG: Making its shell. That's a good way to look at it, that's right.

SH: You've said that an early influence for you was A.N. Whitehead. What did he contribute to your thinking when you were a young scientist?

BG: Whitehead's philosophy is of course that everything that is creative and alive and that the world is a form of organism. So an electron is a kind of organism. The organism is primary and fundamental. Of course, I like that – it makes sense to me, that there is something fundamental about the organism. It's a dualistic theory. In other words you've got matter and mind both represented in this entity.

SH: Were developmental biologists at the time phenomenologists because they focused on observing and recording how organisms develop rather than theorising about this?

BG: I think that would be a good way to put it, yes; they actually observed the extraordinary capacity organisms have for forming wholes out of parts; and of course this is the regenerative capacity; you take a bit of a hydroid and it will regenerate the whole. And so there was a sense in which every single part of a hydroid has this organisational capacity within it; so it can organise itself, self-organise into a body, and so developmental biologists have this tradition of recognising that there is something in the way of organising principles within the organism that go beyond any simple description of genetic programmes, although genetic programmes are constituent elements of the process. But then the question is: how you develop this coherence? How is the organism such a coherent, integrated being? Of course there was always this tendency for people to say: 'Okay, we're going to resolve this into genes. When we know all the genes involved, we will be able to describe precisely the interactions that are involved in this organising activity.'

Now Gerry Webster and I were a bit sceptical about that because there is a difference between seeing something as an integrated whole and seeing it as a collection of parts that then interact in such a way as to generate new properties as you get in the case of emergent properties. Now with emergent properties, of course, it's very important to recognise that you get unpredictable behaviour from complex systems, but the development of organisms goes beyond that. There is an intrinsic wholeness so that the part contains the whole. It wasn't until I encountered the work of Goethe and Steiner that this really became central to my thinking about the mystery of developmental biology.

SH: How did the work of Goethe help you on your quest?

BG: Well, in the late '70s, early '80s, we met somebody called Philip Kilner who was an anthroposophist; he lived in Forest Row and he became a doctor who made some important contributions to looking at the dynamics of the heart; and Philip offered to come to the University of Sussex while we were doing experiments on the developing embryos of *Xenopus* (SH: the African Clawed Toad). He helped us with a phenomenological approach to what we were doing. And so this was our introduction to Goethean phenomenology, through Philip Kilner.

SH: He actually introduced you to Goethe's approach?

BG: Yes, he did. Just think of the experience for me. There I was. I had, since an undergraduate at McGill, thought that the way to understand organisms is to actually understand the principles of organisation in terms of physics and mathematics. Now Goethe said that is not the way to understand organisms; you'll understand them by observing them as dynamic wholes and you develop this capacity to have a dialogue with them; so you really begin to understand what they're doing as subjects. Now that was radical.

SH: It was very intuitive.

BG: It's utterly intuitive.

SH: So with Goethe there's no theory and there's no looking for underlying mechanisms or principles.

BG: No, that's right.

SH: It's about communion.

BG: It's about communion; and therefore it was a radically different approach. It introduced us to Steiner's way of doing science and of course Steiner's description of Goethean science which I then read; and so to people like Lawrence Edwards who was studying the phenomenology of plant buds. Now when I say phenomenology he was actually using projective geometry to describe how these plant buds change at different times at different seasons of the year. In response to the moon, for example – I mean this is the whole integration of cycles that you get in anthroposophy – and so Lawrence Edwards described these things and it was quite a revelation to me as a mathematician. I'd done projective geometry at Oxford and it had fascinated me, but I didn't know how to use it; and then he said: 'Here's how to use it; use it in a descriptive form'; and therefore it's a kind of phenomenological mathematics, beautiful, absolutely beautiful; and so that was a revelation to me; and I thought: 'How can I use this in biology?'; and it's not easy to find ways of applying these ideas rigorously to an observational, experimental situation. That's what he was doing, he was looking at buds, he was actually photographing them and tracing them, the image, and showing the growth patterns; beautiful work.

SH: So were you trying to fill the gap between Goethe's phenomenology and conventional biology?

BG: I realised that there was some big hiatus here and it was only gradually that I realised that this was the whole subjective approach to the organism; in other words, developing a dialogue with the organism as a living being, a sentient living being, now that's radical in science, absolutely radical and it's not something that I felt I could embrace just like that. So that was quite a slow development. It would bubble up only here at Schumacher College; this is where it really flourished.

PART 3

THEORETICAL BIOLOGY

Complexity and Life

FRITJOF CAPRA

Fritjof Capra, PhD, physicist and systems theorist, is a founding director of the Center for Ecoliteracy in Berkeley, California, and a visiting teacher at Schumacher College. He is the author of several international bestsellers, including The Tao of Physics, The Web of Life *and, most recently,* The Hidden Connections. *www.fritjofcapra.net.*

During the last two decades of the twentieth century, a new understanding of life emerged at the forefront of science. The intellectual tradition of systemic thinking, or 'systems thinking', and the models of living systems developed during the earlier decades of the century, form the conceptual and historical roots of this new scientific understanding of life.

Systemic thinking means thinking in terms of relationships, patterns, processes, and context. Over the past twenty-five years, this scientific tradition was raised to a new level with the development of complexity theory. Technically known as nonlinear dynamics, complexity theory is a new mathematical language and a new set of concepts for describing and modelling complex nonlinear systems. Complexity theory now offers the exciting possibility of developing a unified view of life by integrating life's biological, cognitive, and social dimensions (Capra 2002). In this essay, I shall review the current achievements and status of complexity theory from the perspective of the new scientific understanding of biological life.

Metabolism – the essence of life

Let us begin with the age-old question: what is the essential nature of life in the realm of plants, animals, and microorganisms? To understand the nature of life, it is not enough to understand DNA, proteins, and the other molecular structures that are the building blocks of living organisms, because these structures also exist in dead organisms, e.g. in a dead piece of wood or bone.

The difference between a living organism and a dead organism lies in the basic process of life – in what sages and poets throughout the ages have called the 'breath of life'. In modern scientific language, this process is called metabolism. It is the ceaseless flow of energy and matter through a network of chemical reactions, which enables a living organism to continually generate, repair, and perpetuate itself.

The understanding of metabolism includes two basic aspects. One is the continuous flow of energy and matter. All living systems need energy and food to sustain themselves; and all living systems produce waste. But life has evolved in such a way that organisms form communities, the ecosystems, in which the waste of one species is food for the next, so that matter cycles continually through the ecosystem.

The second aspect of metabolism is the network of chemical reactions that processes the food and forms the biochemical basis of all biological structures, functions, and behaviour. The emphasis here is on 'network'. One of the most important insights of the new scientific understanding of life is the recognition that networks are the basic pattern of organisation of living systems. Ecosystems are organised in terms of food webs, i.e., networks of organisms; organisms are networks of cells, organs, and organ systems; and cells are networks of molecules. The network is a pattern that is common to all life. Wherever we see life, we see networks.

It is important to realise that these living networks are not material structures, like a fishing net or a spider's web. They are *functional* networks, networks of relationships between various processes. In a cell, for example, these processes are chemical reactions between the cell's molecules. In a food web, the processes are processes of feeding, of organisms eating one another. In both cases the network is a nonmaterial pattern of relationships.

Closer examination of these living networks has shown that their key characteristic is that they are self-generating (Capra 1996: 95ff.). In a cell, for example, all the biological structures – the proteins, enzymes, the DNA, the cell membrane, etc. – are continually produced, repaired, and regenerated by the cellular network. Similarly, at the level of a multicellular organism, the bodily cells are continually regenerated and recycled by the organism's metabolic network.

Living networks are self-generating. They continually create, or recreate, themselves by transforming or replacing their components. In this way they undergo continual structural changes while preserving their web-like patterns of organisation.

Nonlinear dynamics

The process of metabolism can be summarised by in terms of the following four key characteristics of biological life:

(1) A living system is materially and energetically open; it needs to take in food and excrete waste to stay alive.
(2) It operates far from equilibrium; there is a continual flow of energy and matter through the system.
(3) It is organisationally closed – a metabolic network bounded by a membrane.
(4) It is self-generating; each component helps to transform and replace other components.

These four characteristics all have one thing in common: they are characteristics of a system whose dynamics and pattern of organisation are nonlinear. Non-equilibrium systems are described mathematically by nonlinear equations; networks are nonlinear, multi-directional, patterns of organisation. This is why complexity theory is so important for understanding living systems. As its technical name 'nonlinear dynamics' indicates, it is a nonlinear mathematical theory.

Nonlinear equations have properties that are strikingly different from those of the linear equations commonly used in science. In a linear differential equation, small changes produce small effects and large effects are obtained by adding up many small changes. Mathematically, this means that the sum of two solutions is again a solution, which makes linear equations relatively easy to solve. They are called 'linear' because they can be represented on a graph by a straight line. Nonlinear equations, by contrast, are represented by graphs that are curved, are very difficult to solve, and display a host of unusual properties.

In science, until recently, we always avoided the study of nonlinear systems because of the mathematical difficulties involved in the equations describing them.

Whenever nonlinear equations appeared, they were replaced by linear approximations. Instead of describing the phenomena in their full complexity, the equations of classical science deal with *small* oscillations, *shallow* waves, *small* changes of temperature, and so on, for which linear equations can be formulated. This became such a habit that most scientists and engineers came to believe that virtually all natural phenomena could be described by linear equations.

The decisive change over the last twenty-five years has been to recognise the importance of nonlinear phenomena, and to develop mathematical techniques for solving nonlinear equations. The use of computers has played a crucial role in this development. With the help of powerful, high-speed computers, mathematicians are now able to solve complex equations that had previously been intractable. In doing so, they have devised a number of techniques, a new kind of mathematical language that revealed very surprising patterns underneath the seemingly chaotic behaviour of nonlinear systems, an underlying order beneath the seeming chaos (Stewart 1997).

Let me now review some of the main features of nonlinear dynamics, the theory of complexity (Capra 1996: 112ff.; Mosekilde *et al.* 1988). When you solve a nonlinear equation with these new mathematical techniques, the result is not a formula but a visual shape, a pattern traced by the computer, known as an 'attractor'. An attractor is a geometrical figure in two, three, or more dimensions that represent the variables needed to describe the system. These dimensions form a mathematical space called 'phase space'. Each point in phase space is determined by the values of the system's variables, which in turn completely determine the state of the system.

In other words, each point in phase space represents the system in a particular state. As the system changes, the point representing it traces out a trajectory that represents the dynamics of the system. The attractor, then, is the pattern of this trajectory in phase space. It is called 'attractor', because it represents the system's long-term dynamics. A nonlinear system will typically move in a variety of ways in the beginning, depending on how it is started it off, but then will settle down to a characteristic long-term behaviour, represented by the attractor. Metaphorically speaking, the trajectory is 'attracted' to this pattern whatever its starting point may have been.

Over the past twenty years, scientists and mathematicians explored a wide variety of complex systems. In case after case they would set up nonlinear equations and have computers trace out the solutions as trajectories in phase space. To their great surprise, these researchers discovered that there is a very limited number of different attractors. Their shapes can be classified topologically, and the general dynamic properties of a system can be deduced from the shape of its attractor.

The analysis of nonlinear systems in terms of the topological features of their attractors is known as 'qualitative analysis'. A nonlinear system can have several attractors, and they may be of several different types. All trajectories starting within a certain region of phase space will lead sooner or later to the same attractor. This region is called the 'basin of attraction' of that attractor. Thus the phase space of a nonlinear system is partitioned into several basins of attraction, each embedding its separate attractor.

When we try to assess the achievements and current status of complexity theory, we need to remember, first of all, that nonlinear dynamics is not a scientific theory, in the sense of an empirically based analysis of natural or social phenomena. It is a *mathematical* theory, i.e. a body of mathematical concepts and techniques for the description of nonlinear systems. The most important achievement of nonlinear dynamics, in my view, is to provide the appropriate language for dealing with nonlinear systems. The key concepts of this language – chaos, attractors, fractals, qualitative analysis, etc. – did not exist twenty-five years ago. Now we know what kinds of questions to ask when we deal with nonlinear systems.

Having the appropriate mathematical language does not mean that we know how to construct a mathematical model in a particular case. We need to simplify a highly complex system by choosing a few relevant variables, and then we need to set up the proper equations to interconnect these variables. This is the interplay between science and mathematics. So, the creation of a new language is, in my view, the overall achievement of nonlinear dynamics; and then there are partial achievements in various fields. Among them I shall now concentrate on those achievements that have led to major breakthroughs in our understanding of biological life.

Theory of dissipative structures

The Russian-born chemist and Nobel Laureate Ilya Prigogine was one of the first to use nonlinear dynamics to explore basic properties of living systems. What intrigued Prigogine most was that living organisms are able to maintain their life processes under conditions of nonequilibrium. During the 1960s, he became fascinated by systems far from equilibrium and began a detailed investigation to find out under exactly what conditions non-equilibrium situations may be stable.

The crucial breakthrough occurred, when he realised that systems far from equilibrium must be described by nonlinear equations. The clear recognition of this link between 'far from equilibrium' and 'nonlinearity' opened an avenue of research for Prigogine that would culminate a decade later in his theory of dissipative structures, formulated in the language of nonlinear dynamics (Prigogine & Glasdorff 1971; Capra 1996: 172ff.).

A living organism is an open system that maintains itself in a state far from equilibrium, and yet is stable: the same overall structure is maintained in spite of an ongoing flow and change of components. Prigogine called the open systems described by his theory 'dissipative structures' to emphasise this close interplay between structure on the one hand and flow and change (or dissipation) on the other. The farther a dissipative structure is from equilibrium, the greater is its complexity and the higher is the degree of nonlinearity in the mathematical equations describing it.

The dynamics of these dissipative structures specifically include the spontaneous emergence of new forms of order. When the flow of energy increases, the system may encounter a point of instability, or bifurcation point, at which it can branch off into an entirely new state where new structures and new forms of order may emerge.

This spontaneous emergence of order at critical points of instability, often simply referred to as 'emergence', is one of the most important concepts of the new understanding of life. Emergence is one of the hallmarks of life. It has been recognised as the dynamic origin of development, learning, and evolution. In other words, creativity – the generation of new forms – is a key property of all living systems. And since emergence is an integral part of the dynamics of open systems, this means that open systems develop and evolve. Life constantly reaches out into novelty.

The theory of dissipative structures explains not only the spontaneous emergence of order, but also helps us to define complexity. Whereas traditionally the study of complexity has been a study of complex structures, the focus is now shifting from the structures to the processes of their emergence. For example, instead of defining the complexity of an organism in terms of the number of its different cell types, as biologists often do, we can define it as the number of bifurcations the embryo goes through in the organism's development. Accordingly, the British biologist Brian Goodwin (1998) speaks of 'morphological complexity'.

Cell development

The theory of emergence, known technically as bifurcation theory, has been studied extensively by mathematicians and scientists, among them the American biologist Stuart Kauffman. Kauffman used nonlinear dynamics to construct binary models of genetic networks and was remarkably successful in predicting some key features of cell differentiation (Kauffman 1993; Kauffman 1991; Capra 1996: 194ff.).

A binary network, also known as Boolean network, consists of nodes capable of two distinct values, conventionally labelled ON and OFF. The nodes are coupled to one another in such a way that the value of each node is determined by the prior values of neighbouring nodes according to some 'switching rule'.

When Kauffman studied genetic networks, he noticed that each gene in the genome is directly regulated by only a few other genes, and he also knew that genes are turned on and off in response to specific signals. In other words, genes do not simply act; they must be activated. Molecular biologists speak of patterns of gene expression.

This gave Kauffman the idea of modelling genetic networks and patterns of gene expression in terms of binary networks with certain switching rules. The succession of ON-OFF states in these models is associated with a trajectory in phase space and is classified in terms of different types of attractors.

Extensive examination of a wide variety of complex binary networks has shown that they exhibit three broad regimes of behaviour: an ordered regime with frozen components (i.e. nodes that remain either ON or OFF), a chaotic regime with no frozen components (i.e. nodes switching back and forth between ON and OFF), and a boundary region between order and chaos where frozen components just begin to change.

Kauffman's central hypothesis is that living systems exist in that boundary region near the so-called 'edge of chaos'. He believes that natural selection may favour and sustain living systems at the edge of chaos, because these may be best able to coordinate complex and flexible behaviour. To test his hypothesis, Kauffman applied his model to the genetic networks in living organisms and was able to derive from it several surprising and rather accurate predictions.

In terms of complexity theory, the development of an organism is characterised by a series of bifurcations, each corresponding to a new cell type. Each cell type corresponds to a different pattern of gene expression, and hence to a different attractor. Now, the human genome contains between 30,000 and 100,000 genes. In a binary network of that size, the possibilities of different patterns of gene expression are astronomical. However, Kauffman could show that at the *edge of chaos* the number of attractors in such a network is approximately equal to the square root of the number of its elements. Therefore, the human network of genes should express itself in approximately 245 different cell types. This number comes remarkably close to the 254 distinct cell types identified in humans.

Kauffman also tested his attractor model with predictions of the number of cell types for various other species, and again the agreement with the actual numbers observed was very good. Another prediction of Kauffman's attractor model concerns the stability of cell types. Since the frozen core of the binary network is identical for all attractors, all cell types in an organism should express mostly the same set of genes and should differ by the expressions of only a small percentage of genes. This is indeed the case for all living organisms.

In view of the fact that these binary models of genetic networks are quite crude, and that Kauffman's predictions are derived from the models' very general features, the agreement with the observed data must be seen as a remarkable success of nonlinear dynamics.

Morphology

A very rich and promising area for complexity theory in biology is the study of the origin of biological form, known as morphology. This is a field of study that was very lively in the eighteenth century, but then was eclipsed by the mechanistic approach to biology, until it made a comeback very recently with the emphasis of nonlinear dynamics on patterns and shapes.

A key insight of the new understanding of life has been that biological forms are not determined by a 'genetic blueprint', but are emergent properties of an entire epigenetic network of metabolic processes.

To understand the emergence of biological form, we need to understand not only the genetic structures and the cell's biochemistry, but also the complex dynamics that unfold when the epigenetic network encounters the physical and chemical constraints of its environment. In this encounter, the interactions between the organism's physical and chemical variables are highly complex and can be represented in simplified models by nonlinear equations. The solutions of these equations, represented by a limited number of attractors, correspond to the limited number of possible biological forms. This technique has been applied to a variety of biological forms, from branching patterns of plants and the colouring of sea shells to the nest building of termites (Solé & Goodwin 2000; Stewart 1998).

An good example is the work of Brian Goodwin (1994: 77ff.), who used nonlinear dynamics to model the stages of development of a single-celled Mediterranean alga, called *Acetabularia*, which forms beautiful little 'parasol' caps. Like the cells of all plants and animals, the cell of this alga is shaped and sustained by its cytoskeleton, a complex and intricate structure of protein filaments. The cytoskeleton is subject to various mechanical stresses, and it turns out that a key influence on its mechanical state – its rigidity or softness – is the calcium concentration in the cell. The cytoskeleton is anchored to the cell wall, and its behaviour under the mechanical stresses, therefore, gives rise to the alga's biological form.

Since the mechanical properties of the cytoskeleton at the molecular level are far too complex to be described mathematically, Goodwin and his colleagues approximated them by a continuous field, known in physics as a stress-tensor field. They were then able to set up nonlinear equations that interrelate patterns of calcium concentration in the alga's cell fluid with the mechanical properties of the cell walls.

These equations contain numerous parameters, such as the diffusion constant for calcium, the resistance of the cytoskeleton to deformation, and so on. In nature, these quantities are determined genetically and change from species to species, so that different species produce different biological forms.

Goodwin and his colleagues proceeded to try out a variety of parameters in computer simulations to explore the types of form that a developing alga could produce. They succeeded in simulating a whole sequence of structures that appear in the alga's development of its characteristic stalk and parasol. These forms emerged as successive bifurcations of the attractors representing the interplay between patterns of calcium and mechanical strain.

The lesson to be learned from these models of plant morphology is that biological form emerges from the nonlinear dynamics of the organism's epigenetic network as it interacts with the physical constraints of its molecular structures. The genes do not provide a blueprint for biological forms. They provide the initial conditions that determine which kind of dynamics – or, mathematically, which kind of attractors – will appear in a given species. In this way genes stabilise the emergence of biological form.

Developmental stability

From the origin of biological form, I shall now turn to the development of an embryo (Capra 2002: pp.152–53). Complexity theory may shed new light on an intriguing property of biological development that was discovered almost a hundred years ago by the German embryologist Hans Driesch. With a series of careful experiments on sea urchin eggs, Driesch showed that he could destroy several cells in the very early stages of the embryo, and it would still grow into a full, mature sea urchin. Similarly, more recent genetic experiments have shown that 'knocking out' single genes, even when they were thought to be essential, had very little effect on the functioning of the organism.

This very remarkable stability and robustness of biological development means that an embryo may start from different initial stages – for example, if single genes or entire cells are destroyed accidentally – but will nevertheless reach the same mature form that is characteristic of its species. The question is, what keeps development on track?

There is an emerging consensus among genetic researchers that this robustness indicates a redundancy in genetic and metabolic pathways. It seems that cells maintain multiple pathways for the production of essential cellular structures and the support of essential metabolic processes. This redundancy ensures not only the remarkable stability of biological development but also great flexibility and adaptability to unexpected environmental changes. Genetic and metabolic redundancy may be seen, perhaps, as the equivalent of biodiversity in ecosystems. It seems that life has evolved ample diversity and redundancy at all levels of complexity.

The observation of genetic redundancy is in stark contradiction to genetic determinism, and in particular to the metaphor of the 'selfish gene' proposed by the British biologist Richard Dawkins (1976). According to Dawkins, genes behave as if they were selfish by constantly competing, via the organisms they produce, to leave more copies of themselves.[1] From this reductionist perspective, the widespread existence of redundant genes makes no evolutionary sense. From a systemic point of view, by contrast, we recognise that natural selection operates not on individual genes but on the organism's patterns of self-organisation. In other words, what is selected by nature is not the individual gene but the endurance of the organism's life cycle.

Now, the existence of multiple pathways is an essential property of all networks; it may even be seen as the defining characteristic of a network. It is therefore not surprising that complexity theory, which is eminently suited to the analysis of networks, should contribute important insights into the nature of developmental stability.

In the language of nonlinear dynamics, the process of biological development is seen as a continuous unfolding of a nonlinear system as the embryo forms out of an extended domain of cells.[2] This 'cell sheet' has certain dynamical properties that give rise to a sequence of deformations and foldings as the embryo emerges. The entire process can be represented mathematically by a trajectory in phase space

---

[1] See Goodwin (1994: 29ff.) for a critical discussion of the 'selfish gene' metaphor.
[2] I am grateful to Brian Goodwin for clarifying discussions on this subject.

moving inside a basin of attraction toward an attractor that describes the functioning of the organism in its stable adult form.

A characteristic property of complex nonlinear systems is that they display a certain 'structural stability'. A basin of attraction can be disturbed or deformed without changing the system's basic characteristics. In the case of a developing embryo this means that the initial conditions of the process can be changed to some extent without seriously disturbing development as a whole. Thus developmental stability, which seems quite mysterious from the perspective of genetic determinism, is recognised as a consequence of a very basic property of complex nonlinear systems.

Origin of life

The last example of my review of applications of complexity theory to problems in biology is not about an actual achievement but about the potential for a major breakthrough in solving an old scientific puzzle – the question of the origin of life on Earth (Capra 2002: 17ff.).

Ever since Darwin, scientists have debated the likelihood of life emerging from a primordial 'chemical soup' that formed four billion years ago when the planet cooled off and the primeval oceans expanded. The idea that small molecules should assemble spontaneously into structures of ever-increasing complexity runs counter to all conventional experience with simple chemical systems. Many scientists have therefore argued that the odds of such prebiotic evolution are vanishingly small; or, alternatively, that there must have been an extraordinary triggering event, such as a seeding of the Earth with macromolecules by meteorites.

Today, our starting position for resolving this puzzle is radically different. Scientists working in this field have come to recognise that the flaw of the conventional argument lies in the idea that life must have emerged out of a chemical soup through progressive increase of molecular complexity. The new thinking begins from the hypothesis that very early on, *before* the increase of molecular complexity, certain molecules assembled into primitive membranes that spontaneously formed closed bubbles, and that the evolution of molecular complexity took place inside these bubbles, rather than in a structureless chemical soup.

It turns out that small bubbles, known to chemists as vesicles, form spontaneously when there is a mixture of oil and water, as we can easily observe when we put oil and water together and shake the mixture. Indeed, the Italian chemist Pier Luigi Luisi (1996) and his colleagues at the Swiss Federal Institute of Technology have repeatedly prepared appropriate 'water-and-soap' environments in which vesicles with primitive membranes, made of fatty substances known as lipids, formed spontaneously.

The biologist Harold Morowitz (1992) has developed a detailed scenario for prebiotic evolution along these lines. He points out that the formation of membrane-bounded vesicles in the primeval oceans created two different environments — an outside and an inside — in which compositional differences could develop. The internal volume of a vesicle provides a closed micro-environment in which directed chemical reactions can occur, which means that molecules that are normally rare may be formed in great quantities. These molecules include in particular the building blocks of the membrane itself, which become incorporated into the existing membrane, so that the whole membrane area increases. At some point in this growth process the stabilising forces are no longer able to maintain the membrane's integrity, and the vesicle breaks up into two or more smaller bubbles.

These processes of growth and replication will occur only if there is a flow of energy and matter through the membrane. Morowitz describes a plausible scenario of how this might have happened. The vesicle membranes are semi-permeable, and thus various small molecules can enter the bubbles or be incorporated into the membrane. Among those will be so-called chromophores, molecules that absorb sunlight. Their presence creates electric potentials across the membrane, and thus the vesicle becomes a device that converts light energy into electric potential energy. Once this system of energy conversion is in place, it becomes possible for a continuous flow of energy to drive the chemical processes inside the vesicle.

At this point we see that two defining characteristics of cellular life are manifest in rudimentary form in these primitive membrane-bounded bubbles. The vesicles are open systems, subject to continual flows of energy and matter, while their interiors are relatively closed spaces in which networks of chemical reactions are likely to develop. We can recognise these two properties as the roots of living networks and their dissipative structures.

Now the stage is set for prebiotic evolution. In a large population of vesicles there will be many differences in their chemical properties and structural components. If these differences persist when the bubbles divide, we can speak of 'species' of vesicles, and since these species will compete for energy and various molecules from their environment, a kind of Darwinian dynamics of competition and natural selection will take place, in which molecular accidents may be amplified and selected for their 'evolutionary' advantages.

Thus we see that a variety of purely physical and chemical mechanisms provides the membrane-bounded vesicles with the potential to 'evolve' through natural selection into complex, self-producing structures without enzymes or genes in these early stages. A dramatic increase in molecular complexity must have occurred when catalysts, based on nitrogen, entered the system, because catalysts create complex chemical networks by interlinking different reactions. Once this happens, the entire nonlinear dynamics of networks, including the spontaneous emergence of new forms of order, comes into play.

The final step in the emergence of life was the evolution of proteins, nucleic acids, and the genetic code. At present, the details of this stage are still quite mysterious. However, we need to remember that the evolution of catalytic networks within the closed spaces of the protocells created a new type of network chemistry that is still very poorly understood. This is where complexity theory could lead to decisive new insights. We can expect that the application of nonlinear dynamics to these complex chemical networks will shed considerable light on the last phase of prebiotic evolution.

Indeed, Morowitz points out that the analysis of the chemical pathways from small molecules to amino acids reveals an extraordinary set of correlations that seem to suggest, as he puts it, a 'deep network logic' in the development of the genetic code. The future understanding of this network logic may become one of the greatest achievements of complexity theory in biology.

The Language of Living Processes

PHILIP FRANSES

Philip Franses studied mathematics at Oxford University, working with creative software systems. He did further studies with Brian Goodwin at Schumacher College where he is now teaching complexity and chaos theory on the MSc in Holistic Science that Brian set up with Stephan Harding.

I had the opportunity to work with Brian at Schumacher College from 2005 when arriving to do the MSc until his death in 2009. We worked together on the way words take on meaning according to how context resolves ambiguities creatively. This work applied to biology is summarised here.

Abstract

In thermodynamics a direct correlation is made between the freedom of the atoms (or entropy, disorder) to move around, and the energy or temperature of the gas as the whole. This translates directly into living systems, where the scope of possibilities at a micro level fuel the unity of the organism at a macro level, giving internal resource for the organism to exploit its fluidity of response in the world. On the other hand the living system is constrained from falling into a state of total disorder or maximum freedom of parts by the steady supply of energy gleaned from outside.

In my work with Brian Goodwin on a thesis 'Living Ambiguity' in 2006 at Schumacher College and thereafter, we concentrated on a computer model of language, first suggested by Cancho and Solé.

What is the optimum balance of existing in freedom of possibilities that neither disintegrates into a will-less anarchy nor resorts to a fixed indexical one-one relation of parts to whole functions, but makes available an internal fluidity as resource to the whole? The study characterises biological development by the signs in which individual processes communicate their searching for a common language of realisation.

Language

Brian often referred to Waddington's conclusion in Volume 4 of the series *Towards a Theoretical Biology* in 1972. At the end of the Epilogue, Waddington suggested that language in the sense of a communication process that involves instructions for effective action 'may become a paradigm for the theory of General Biology' (Waddington 1972, p.289).

The development of the biology of language, or biosemiotics, has developed tremendously in the intervening years (Favareau, Hoffmeyer). The very notion of genetic code as pointing to something more than a material trigger for cellular activity has become more and more pertinent in the discoveries of molecular biology. However huge puzzles remain in how such a coordination of sign-potentials should be founded, whether as a layer on top of the conceptual description of the organism, or whether physical properties of organisms emerge from a communication in signs. The whole signs its nature in the language of its becoming.

An example of seeing the world from the question of how the whole is assembled is given in physics. Thermodynamics pulls apart the gas into the energetic motion of its molecules as parts, to construct from their internal freedom a notion of the temperature of the gas as a whole. In seeing the dynamic relation of parts to the whole, internal properties of the whole are discovered additional to the external causes acting on the parts as separate entities. There is a particular energy available in the relatedness of the parts, and their dependence on the creating of the whole.

If one simply sees the molecules making up a gas as entities already isolated in their separate distinction, then one would miss the vital relationship of their related freedoms that gives to the whole an internal energy. In the same way a subtle relation between, say, the cells in their separate distinctions, makes available an internal energy to the organism as a whole, which it can utilise as a potential resource.

Our hypothesis is that such internal dynamics provides a resource for the organism to focus its own future in a similar way that language frames a meaning from its deft use of the word parts. 'New notions, analogous to thermodynamic quantities are directly applicable to the properties of biological systems, developed on the basis of a dynamic theory of molecular control processes'. (Goodwin 1963, p. vii)

Our inquiry is based on a computer model of language that uncovers this primary role of ambiguity or freedom in language. We apply this model to living systems in general.

Model

Language has enormous entropy, or possibility for disorder, since it is the coordinated and appropriate choice of words that gives greater meaning than the sum provided by the individual units. When used skilfully, each word is perfectly optimised and fitting its service in expression of the whole sentiment. There is a generic property of language that is expressed in terms of word frequencies, known as Zipf's Law. This originally stated that there is a self-scaling or power law relation between the frequency of words used and their rank order in the list of all words ordered by frequency. Zipf's Law has been shown to be true for all natural languages. If one counts the most common word in the English language, 'the', it will be twice as common as the next most common word, 'and', three times as common as the third most common word, 'of', etc.

The interpretation that we present here as a basis for exploring complex aptitude for response in living process comes from a model (Cancho & Solé 2003) with respect to the evolution of language. Brian and I were able to confirm their findings, on various sets of word-object representations (Franses 2006, p.33–38).

The model investigates the influence of ambiguity on a simple model of words applying to objects. It explores the tension between the entropy (disorder) of informational possibilities and the negative entropy (order) of their unique signification. The hypothesis is that ambiguity creates channels for information that instead of residing passively in the physical system are explored actively for their ability to be integrated into coherent meaning.

In the model, a number of words are associated randomly with a number of objects. Entropy as a quality of disorder has a perfectly precise mathematical quantification as the logarithm of the possibilities open to each word-element.

Entropy and negative entropy are interpreted in the model to relate to speaker effort and hearer effort of using words to convey objects, where words may refer to more than one object

The speaker effort decreases with the number of different objects to which the same word can refer. If one word is used for all objects, the speaker effort is lowest but the system will be highly ambiguous and communication will be impossible. (The more disordered the language the less the speaker has to impose order). On the other hand the speaker effort will be highest when each word is associated with one particular object (the speaker will have to say the appropriate word).

The hearer effort is to find a path to resolve the numerous possibilities of the words into a particular contextual meaning. The hearer effort is highest when there is the most ambiguity. In the case of using one word for all objects it is impossible to arrive at a logical path to the identity of the objects. On the other hand in a one to one association of words to objects, no effort is required to unravel the meaning of what is said, so hearer effort is lowest. The logical path to the revelation of the object inferred by the words is obvious.

Thus speaker and hearer effort are in mutual tension. A network will resolve itself according to the relative emphasis that is put on the efforts.

1. If there is little importance in how a feature expresses itself, the speaker effort will be minimised and there will be a highly variable anarchic system in which the behaviour of elements pays no attention to other elements.
2. If it is crucial how order is interpreted for other parts of the system, then the hearer effort will be minimised and there will be a highly deterministic system in which each element has a fixed meaning. This is machine language that requires precision but allows no creative sensitivity to context.
3. Where the emphasis is evenly distributed and both

efforts have to be accommodated, a compromise solution is found. The minimum effort, instead of representing an extreme example of one type of organisation, lowers the threshold that has to be crossed from the side of both speaking and hearing.

A power law of the same form as Zipf's Law for languages, emerges as the optimum distribution of words to objects that realises the path of least connection spanning the threshold. 'In language, the power law distribution of words is a necessary condition for a deeper and highly significant property of natural languages: the ambiguity of meaning in utterances'. (Goodwin 2007, p.107) The solution requires creative interpretation of ambiguity to make sense of the system according to the context, while ensuring a navigable path crossing from words to meaning. This corresponds to the state of creative language in which ambiguity is allowed and is freely interpreted into meaning by the skill of the listener.

A power law states that some word will have ambiguous association with a large number of other objects, the next ranked word an ambiguity with a half of the objects, the next ranked word a third of the objects, etc. A large amount of words will be unambiguous.

Interpretation

The model illustrates that for the genetic code to have maximum efficacy, then it should be seen as signs whose interpretation can be freely made by context. Any literal description of the organism in terms of exactly interpretable material triggers is going to be inefficient and inflexible. What is more the power law gives a tell-tale expression of an organisation of signs.

In *Laws of Form*, Spencer Brown exchanges a logic of concepts of what is already formed with the activity by which things become distinct, effecting and effected by those other processes with which it has to form a common order:

> A theorem is no more proved by logic and computations than a sonnet is written by grammar and rhetoric, or that a sonata is composed by harmony and counterpoint, or

a picture painted by balance and perspective. Applied science is seen as drawing sustenance from a process of creation, with which it can combine to give structure, but which it cannot appropriate. (Spencer Brown 1969, pp.82–83)

A conventional biological analysis works by pulling apart the world into distinct conceptual units of genes, proteins and cells; and then to reintegrate the units into a comprehensible logic of interactions.

A sign on the other hand is an indicator of somewhere to where one is going, with no other information existing from the causal past able to prepare one for such an eventuality. A sign is the opposite of a concept that roots the world on the basis of its origin; a sign is giving to the world of the future an indication out of the completing nature of its process.

A sign communicates a journey giving partial information on addressing a potential that is still in the future. In this way the biology of signs, known as biosemiotics, explores how different actors exchange information about context, steering individual journeys to an economy of effective arrival. The language of signs is able to navigate through the landscape of interactions gathering the information that relates an individual task with the context of the whole.

The biology of signs addresses the world in its potential to distil the content of environment into an indication of direction. Instead of the concept imposing itself on the world around, the sign orientates a common direction. The sign in relation to the concept reflects the symmetry of future to past.

A symbolic representation instead of seeking to capture exactly the authority of an object, seeks to share a partial realisation of journey to instruct a collective economy of arrival at a destination.

Legacy

Many of Brian's key discoveries in biology, say around the dynamic nature of the embryo, had come from an intuitive understanding of the symbolic significance of the path of development. This intuitive understanding has often been vindicated in later experimentation. Brian saw into the language of signs by which an organism developed itself.

Brian's phrase 'Minimum Effort, Maximum Grace', summed up the conviction that the discovery of the symbolic dynamic of an organism, was an optimum description of how the organism created the distinct boundary of its being. The symbolic dynamic was not draining the organism into the wisdom of a dry explanation, the representation of the environment by the organism, followed a subtle logic in an active dynamic.

It was at Schumacher College with its multi-discipline approach, that Brian could truly explore symbolic understanding as a basic principle of how the organism optimises its own powers of representation to make the most of its potential. Brian had opened to the secrets of a single celled organism or the working of the heart, to assert that the economy of mathematical description had something to say about the organisms themselves, in their act of assembling their parts to express the whole. The organisms told him their living secrets and it was a misrepresentation of the logic to portray this as a static value-free causal connection. What Brian wanted the opportunity to explore was a new model of logic that could allow the influence of the organism itself to arrive at its form in the dynamic of its arising. To do this he had to step right outside the normal scientific logic, where the explanation is equivalent to the thing described.

One of the first found supports in his search was the German poet Goethe (1749–1832), also a formidable scientist. Together with Brian, Goethe's work was given modern foundation by two colleagues: Henri Bortoft, linking Goethe's methodology to the phenomenologist tradition in continental Europe; and Margaret Colquhoun, who explored Goethe's practical methodology as applied to plants and landscape.

More than genes

Goethe recorded a methodology of approaching the plant to learn the subtle relation of whole: part from the plant itself. Here Brian found an accurate foundation to his own experience of learning from nature, in a living encounter. The reinstatement of this living source meant one would have to re-vivify how parts and whole related, as a new subject of the investigation.

An example of Goethe's work is the study of the plant. Goethe's methodology of seeing works to recognise the signs of the plant's dynamic of growing and to follow these signs to a living realisation of the whole nature of the plant. The plant is first studied in a classical manner, recording and noting its various structures, using an objective approach. However the facts gleaned from this exercise are not then turned into an abstract conceptualisation of the life of the plant. Instead the observations are treated as signs pointing to a common process of development of the plant taking it from seed to maturity. The signs have to be put together into the imagination, so that their destination – the real full nature of the whole plant – suggests itself. The following of the signs gives an intuitive intimation of the dynamic fulfilment the plant gives to the process of its growth.

By this process Goethe discovered a continuous transformation in the various organs of leaf, sepals, petals, stamens and carpels adjusted to the related nuance of the function of each in the development of the plant. The plant can select in context how the environment requires distinction to manifest a particular version of this organ. It can thus be that in different conditions of climate and environment, the quality of making distinctions results in the plant appearing differently.

As Goethe himself observed when travelling through the Alps on a voyage from his native Germany (Goethe 1787), many Alpine versions of plants he knew at home had subtle variations. From this he deduced that the act of making the distinction into the various organs could translate itself differently according to circumstance, there was no fixed elemental basis out of which the plant constructed itself. For the difference in the alpine and the native variations held to a harmony that was in each case different through all its parts. It was a living interpretation that appeared in two quite different though clearly associated varieties.

The act of distinction is something that is quite obvious to us in language. Words suggest themselves as signs that capture some part of what one is expressing, pulling together into representing the whole sentiment one is trying to articulate. The twofold act of distinction allows one to think up the words for the pertinence of their individual meaning, as well as embed these into a context of the sentence, whose grammatical rules integrate the words into the text. Similarly the cell or the organ is able to benefit from the twofold act of distinction, to work on the particular quality of its function within its integration in the whole.

To test this, we looked with Professor David Murphy and Dr Charles Hindmarch at Bristol University at some RNA microarray data they had gathered (Franses 2007, Hindmarch 2006). The RNA gives a profile of the genes expressed with their intensity as the organisms respond to various situations. The inquiry validated the results of Lu that, 'Recent investigations of data in gene expression databases for varying organisms and tissues have shown that the majority of expressed genes exhibit a power-law distribution with an exponent close to -1 (i.e. obey Zipf's law)' (Lu 2005).

The traditional inquiry sees the words of the genetic code as a basic resource the organism possesses and attempts to map this vocabulary onto a rigid interpretation of the making of proteins that then instructs the cells in their functions. Murphy and Hindmarch's detailed research had focussed on the role of individual genes in their relation to proteins and the behaviour of the organism. However as more and more data became available, it became apparent that there was no simple formula by which one could isolate one gene and its associated protein to a behavioural change. Our suggestion was that the genes were as signs that were read and informed by the context of cells in putting together a response to environmental change.

Searching for the more generic order over the whole data-set, revealed how the RNA very loosely obeyed a Zipf Law type relation of their intensities to their ranks. Our interpretation of this was that the map of the organism's response was being put together dynamically from the signs of genetic pointers to the whole behaviour.

The genes are activated according to a preparatory assessment of their place, signposting the choices that are made in selection of the integrative response of the whole organism. The logic of the genes is one of creating distinctions that are advantageous in the construction of the whole.

If we return to Goethe's ambiguous distinction of the organs, various genes were later discovered that actually switched the organ's expression between its various forms:

> Goethe was right when he proposed that flowers are modified leaves. It seems that four genes involved in plant development must be expressed together to turn leaves into floral organs. According to this model, the identity of the different floral organs – sepals, petals, stamens and

carpels – is determined by four combinations of floral homeotic proteins known as MADS-box proteins.

The protein quartets, which are transcription factors, may operate by binding to the promoter regions of target genes, which they activate or repress as appropriate for the development of the different floral organs. (Theissen & Saedler 2001)

The genes in this case are those subtle prefixes of linguistic precision that switch the meaning of the whole organism along its various paths. The genes are as signals that set the tracks down different destinations, mere semantic triggers that can be pulled at the last minute to turn the whole upon a different route. The genes distinguish in their setting, a vital differentiation for the process of disambiguation that is already under way when the trigger is activated in response to context.

The signs of genes communicate an informational potential that gives direction to the various processes in the organism to complete coherently.

What is true at the level of the genes can be applied to the whole organism. Since there is available to the organism many more possibilities than are actualised in a given situation, there must be some process that steps from the world of internal possibilities to the world of integrated actions in which the organism functions.

Representation

Consider the blackbird. As one watches it skip across the ground, checking its way with each step, and heading towards the leaves to pick for a worm, what is the world in which it moves? The being of the blackbird, its instincts, its individuality all are present in its representation of the world that interprets the information from the eye that turns, head cocked, to meet one's watchful gaze. There are two aspects to the relation of individual to world:

—the representation of the world of the blackbird is given additional colour by its instincts, its blackbird-ness, its circumstance, etc;
—functionally the blackbird is tested as to how the

outcome of its behavioural choices equip the bird for its place in the physical world.

This ambiguity between the represented world and the actual world is what gives the vitality to the blackbird. The enormous freedom in possibilities between how this representation of the world translates into the actuality of its behaviour gives the blackbird an internal resource to respond energetically. In this possibility, the blackbird steps out from the mechanical causal backdrop of the physical world in which it lives.

Just like a gas has thermodynamic energy in the possibilities in the freedom of its atoms, so the organism has energy in the freedom of how it translates its representation into actions.

The difference between the representational and the causal linear logic of the actualised, gives to the awareness of the blackbird the attention that is forever seeking to navigate its place in the world through the medium of its own possibility. The blackbird is alive in the choices that forever allow context to determine the particular of its function. The blackbird knows its environment, exactly because its representation is not privileged to tell it the whole truth of the world. The depth in representational possibilities gives a knowing that is much more than an exact knowledge of reality. The blackbird lives through the signs that open up its possibilities into an interpretation with its environment combining as a lens to focus on specific causal action.

This flexibility in establishing the integrity of being extends beyond the single individual. Consider further bacteria's collective action for sporulation:

> Sporulation is a process executed collectively and beginning only after 'consultation' and assessment of the colonial stress as a whole by the individual bacteria. Simply put, starved cells emit chemical messages to convey their stress. Once all of the colony members have sent out their decisions and read all other messages, sporulation occurs if the majority vote is in favor.
> (Ben Jacob 2004)

The bacteria behave through a sign that is as a junction, indicating a road ahead either to entering a dry spore state to see out a drought period, or to continue in the feeding from an abundant environment. The future they decide upon is a destination that they read through the signs of their fellow bacteria and interpret collectively.

The logic is that of language, where words act out their meanings in context, transparent to the whole they determine in their outcome. The word is uncommitted until finding its meaning, free in what it becomes, without being used up as would some fixed prescription. The word has the lightness to fly through its invocation, to take on the meaning it learns through its journey.

Conclusion

Brian challenged the classical biological dogma reducing biological function to the genetically-driven chemistry of cells. Instead he explored the mathematics of possibility. Complexity theory, chaos theory and the Turing mechanism of activation/inhibition were applied to the self-organising of structure. Simple organisation arose through the focussing of potential, rather than an inheritance of a functional blueprint for behaviour that was exactly followed. Yet still this different mathematical approach failed to capture the flavour of biological development.

Biology happens all at once in various critical moments where potential that has lain dormant, arises suddenly in pivotal expression of collective quality. An embryo cell suddenly voices a directed development out of the homogeneity of being into the fullness of meaning. The excitement of our discovery of primary ambiguity was in exploring some of the ways freedom and order interrelate.

In inverting the conceptual model of biology that fixes existence in terms of what it has been in the past, the biology of signs points forward to actively realising a coherence that is targeted to the future.

Keeping the Gene in its Place

JOHANNES JAEGER AND NICK MONK

Johannes Jaeger started working with Brian during his time as an MSc student in holistic science at Schumacher College. His collaborations and discussions with Brian influenced the direction of his scientific trajectory profoundly.

Nick Monk was encouraged by Brian to change the focus of his research from quantum theory to biology, and continues to benefit from his wisdom and inspiration.

Brian Goodwin was never part of the mainstream. His views and ideas on the development and evolution of organismic form were always radically different from those of the majority of molecular biologists, who often perceived Brian and his theories as reactionary. Many of those who engaged in discussion with him argued that he criticised the predominance of gene-based explanations in biology (and hence the major source of progress in the field at the time, see Keller 2000) without providing any specific, viable alternatives.

This reaction to Brian's work needs to be understood in the context of the prevailing trends in developmental biology during the last forty years of the twentieth century. By the early 1960s, it was clear that biological processes were underpinned by molecular and genetic interactions. This raised two critical questions. First, what is the detailed nature of these interactions, and second, how do the dynamics of organismic form emerge from them? While answering each of these questions is clearly important, the mainstream approach focused almost exclusively on addressing the first. Remarkable and important progress was made, providing a detailed understanding of the molecular and genetic components of developing systems. Sadly, this progress fostered a climate that not only largely ignored the second question – that of emergence – but also tended to dismiss any attempt to address it as misguided.

But times have changed. Biology is currently undergoing a paradigm shift. Instead of single, isolated genes, we are now beginning to study integrated systems – cells, tissues, organisms, even ecological networks as a whole – using computational and mathematical models combined with quantitative measurements. This approach, called systems biology, provides a very different perspective on the nature of biological processes, shifting focus away from genes as independent agents to their embedded role in a complex, dynamic biological context (see, for example, Gilbert & Sarkar 2000; Robert *et al.* 2001; Kitano 2002; Wolkenhauer 2002; Noble 2008; Gatherer 2010). Here we argue that this perspective is strikingly similar to what Brian had been proposing all along. Systems biology now provides quantitative evidence for many of the concepts at the foundation of Brian's theories. Clearly, his ideas had been visionary rather than reactionary. Still, surprisingly few biologists nowadays know of Brian's contributions, which are only too rarely acknowledged in the literature.

In this chapter, we review Brian's thinking on cellular, developmental and evolutionary dynamics, which is both at the origin and the core of his wide-ranging interests in biological systems (see Goodwin 1994; Webster & Goodwin 1996; Solé & Goodwin 2000; Goodwin 2007, for a general overview). We aim to show how many of the concepts and theories he considered crucial – such as the importance of cellular dynamics, morphogenetic or generative fields, and his structuralist theory of rational morphology – are taking centre stage in biology once again, this time backed up by quantitative evidence that was not available when Brian first proposed his ideas.

Cellular dynamics and oscillations

Biological form arises through dynamic regulatory processes during the life cycle of an organism. Such processes consist of interactions between metabolites, ions, genes, cells, and other factors, which result in characteristic patterns of cell differentiation in space and time. One of the best-studied examples of this is the regulation of gene expression during development (see, for example, Davidson 2006; Gilbert 2010). Here, we will define a gene as a DNA sequence in the genome that encodes a RNA or a protein (Figure 1A). These gene products are only produced at specific times, or in specific cell types or tissues, through the processes of transcription and translation (gene expression). The exact pattern of expression depends on proteins, called transcription factors, which bind to the regulatory DNA sequences that determine the expression of the gene. Of course, these transcription factors are themselves encoded by genes that are regulated by other transcription factors, and so on. This results in complex regulatory networks (Figure 1B) containing dozens or hundreds of genes (and/or other factors, as stated above). If we want to understand biological form, we need to understand the temporal and spatial dynamics of the regulatory interactions in such networks.

We will first turn our attention to Brian's earliest work on the temporal organisation in cells (Goodwin 1963; Goodwin 1966; Goodwin & Cohen 1969). It is no coincidence that Brian started out studying temporal cellular dynamics. The fundamental truth that everything in biology is dynamic has always been at the core of his thinking. Nothing ever stays the same. Everything is process.

Of course, this notion by itself was not new. Brian's mentor, Conrad Hal Waddington (1975), was strongly advocating it at the time. However, information on the basic molecular and genetic mechanisms operating within cells was just beginning to emerge from laboratory studies, raising the challenge of explicating the link between cellular and systems-level dynamics. Jacob and Monod's pioneering and hugely influential work on gene regulation in bacteria demonstrated that gene expression could affect the temporal regulatory dynamics of cells (Jacob & Monod 1961; Monod & Jacob 1962). Brian took these ideas and addressed the critical question of how one could account for system-level dynamics in terms of molecular genetics. What was new about Brian's approach was his use of quantitative mathematical modelling and analysis to explore the sometimes counter-intuitive consequences that temporal regulatory processes could have for cell dynamics and pattern formation. Unlike Jacob and Monod's work, Brian's focus wasn't on the molecular details of these systems, but on their potentially far-reaching consequences for our understanding of biological processes in general.

Brian chose to focus on the specific question of the emergence of coherent rhythms in organisms, such as the circadian rhythms that regulate physiological processes and activity in both plants and animals with a 24-hour period. The reasons for this were twofold. The

155

first reason was pragmatic: Brian realised that it was necessary to find a system for which precise quantitative data were available. At the time, biological clocks were the only system for which this was beginning to be the case. The second reason was more profound: Brian argued that oscillatory behaviour was extremely likely to play an important role in determining the behaviour of cells and tissues generally. This view placed central importance on transient cellular *dynamics*, in sharp contrast to the emerging (if implicit) paradigm of development as a series of switch-like transitions between steady states.

Brian identified negative feedback, operating on the elements that control the rate of expression of a gene, as a likely candidate for the

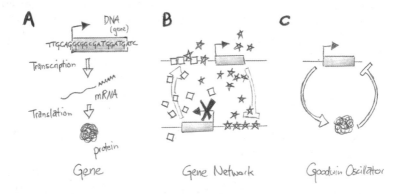

Figure 1: (A) A gene: the coding sequence of the DNA (shaded box) is transcribed into messenger RNA (mRNA), which in turn is translated into protein. DNA sequence is composed of four bases, abbreviated A, C, G, and T.
(B) A gene network: two genes (represented by boxes with associated black arrows indicating transcriptional activity) encode proteins called transcription factors (represented by squares and stars). Such factors bind to regulatory sequences, which often lie upstream of the protein-coding sequence of a gene. They can either have an activating or repressing influence on transcription (represented by white arrow and T-bar, respectively). The transcriptional activity of a gene is determined by the combination of transcription factors (activators and repressors) bound to its regulatory sequences. In this way, genes can switch each other on (upper) or off (lower gene), forming a regulatory network. Regulation of gene expression across time and space by such networks is crucial for pattern formation during development. Note that transcription factors can bind to and regulate their own genes (star-shaped proteins in panel B bind to their own gene).
(C) A Goodwin Oscillator: oscillations in gene expression are based on negative feedback regulation. In the example shown, the protein product of a gene is repressing its own production by inhibiting transcription. Since there is a delay between the initiation of transcription and the appearance of the protein product at the gene's regulatory sequence, this system can show sustained oscillatory behaviour. See text for details.

principal generator of oscillatory behaviour (Figure 1C). Negative feedback was already well known to be widespread in biology. Homeostasis – the maintenance of physiological variables such as body temperature or the concentration of carbon dioxide and glucose in the blood stream of mammals close to a set point – relies on negative feedback control. In contrast to this stabilising role, negative feedback can also generate oscillations. Such oscillations were well known to engineers, but in a biological context were almost exclusively seen as an undesirable feature. Brian took the important step of not only realising that the negative feedbacks present in molecular and genetic interactions could generate important oscillatory dynamics, but also providing a concrete mathematical foundation for their study in terms of enzyme kinetics and the regulation of gene expression (Goodwin 1963; Goodwin 1966; Goodwin & Cohen 1969).

Brian's models helped to initiate, and have inspired, a rich vein of modelling of cellular oscillations (see Goldbeter 1997; Maroto & Monk 2009, for reviews). For many years following Brian's original work, circadian rhythms provided the most compelling example of a cellular oscillator, and are now seen as a paradigm for integrative systems approaches to biology (see, for example, Hogenesch & Ueda 2011). Although remarkable progress has been made in fleshing out the molecular details of the circadian clock, a testament to Brian's insight is the fact that negative feedback remains at the heart of the mechanism (Novak & Tyson 2008).

Brian's prediction that oscillatory dynamics should underpin a wide range of developmental and organismic processes is also turning out to be correct. Recent advances in live imaging of the levels of expression of the molecular constituents of cells and tissues have revealed an ever increasing collection of fundamental oscillatory processes with periods ranging from minutes to several hours (Kruse & Jülicher 2005; Maroto & Monk 2009). These oscillations regulate not only the dynamic responses of individual cells, but also play a central role in pattern formation. A striking example is provided by somitogenesis in vertebrates – the process whereby the basic segmental pattern of the trunk is established during embryonic development (Dequéant & Pourquié 2008). Here, oscillations in gene expression (driven by negative feedback regulation) co-ordinate the periodic assignment of groups of cells to individual tissue segments (Lewis 2008).

These examples illustrate the importance of oscillatory dynamics for regulating the temporal behaviour of cells and for assuring the co-ordinated behaviour of groups of cells in development. A growing body of evidence now supports the idea that feedback-driven oscillations also play a more unexpected role in another central process in development – the diversification and differentiation of cell types.

Theoretical studies by Kaneko and Yomo built on Brian's ideas by exploring the dynamics of a growing population of coupled cells, each of which contained an oscillatory biochemical network (Kaneko & Yomo 1997). Strikingly, they found that after an initial period of cell divisions during which the cells remained synchronous, the population split into locally synchronous groups of cells that had different phases with respect to each other. The oscillatory biochemical networks in the cells drove diversification of an initially homogeneous population, and provided a mechanism for cells to retain their identity and pass it on to daughter cells. This theory of differentiation – called isologous diversification – provided a theoretical realisation of Brian's claim that oscillatory dynamics should play a role in many aspects of development. Further extensions of this work in the light of more recent experimental evidence have extended the idea to stem cells and cancer (Kaneko 2009; Kaneko 2011).

A direct link between Brian's insight into the essential role of oscillatory cellular dynamics has recently emerged from studies of the dynamics of the developing nervous system. In vertebrates, neurons differentiate from a dividing population of progenitor cells over an extended period of development. The rather measured tempo of neural differentiation is important to ensure that an appropriate diversity of neural cell types is generated. Time-lapse imaging in live developing embryos has shown that the biochemical differentiation network in neural progenitor cells goes through an extended oscillatory phase before the cells commit to a neural cell fate, and that these oscillations are necessary to prevent premature differentiation (Shimojo *et al.* 2008). Mathematical models of the neural differentiation network show that feedback of the type proposed by Brian is the critical feature of these networks that supports both oscillation and differentiation at the same time (Monk 2003; Veflingstad *et al.* 2005; Momiji & Monk 2009).

Positional information and pattern formation

While Brian was emphasising temporal aspects of cellular dynamics, his colleague Lewis Wolpert was arguing that such temporal mechanisms alone are not sufficient to understand pattern formation during embryogenesis. He formulated a simple toy model, the French Flag (Figure 2A), to demonstrate why spatial aspects need to be taken into account (Wolpert 1968). In this model, a substance (called a 'morphogen', after Turing 1952) is synthesised and released at one end of a tissue (the source), and diffuses to the other end of the tissue, where it becomes degraded (the sink). This results in a linear gradient of morphogen concentration across the tissue (Figure 2A). Wolpert argued that cells within the tissue have the ability to measure morphogen concentration accurately, and to exhibit distinct concentration-dependent responses. Specifically, cells exposed to concentration ranges lying between discrete thresholds switch on distinct target genes in response to the morphogen. If represented by the colours blue, white, and red, the different sub-territories of gene expression across the tissue form a pattern resembling the French Flag (Figure 2A). In this model, local morphogen concentration encodes 'positional values' that are read by the cells.

This simple model not only explains how a tissue can be patterned by a gradient, but also suggests how the pattern can adapt to differences in tissue size: under the assumption of a linear gradient (as described above), the position of the thresholds scales proportionally with the size of the tissue, according to the law of similar triangles (Figure 2A).

Wolpert generalised his model by proposing that the morphogen could be the same in different biological processes and species, as long as the tissue's reaction to it was specific to its context. In this way, morphogen gradients can be seen as encoding an abstract and universal co-ordinate system imposed on any particular tissue. Wolpert called this co-ordinate system 'positional information' (Wolpert 1969).

The French Flag model was very successful in showing that spatial aspects of regulation are crucial to understand pattern formation. On the other hand, there are three main problems with Wolpert's argument. First, it puts a great (some say excessive) amount of pressure on the cells that measure morphogen concentration (the problem of interpretation). They need to be able to do this very accurately for precise patterning in the presence of molecular fluctuations in the gradient, a problem which was recognised by Wolpert himself (see the appendix in Wolpert 1989). The second problem is that the French Flag model is entirely static and feed-forward. It does not consider dynamic aspects of pattern formation at all. A fixed co-ordinate system is imposed on a tissue, and no feedback regulation occurs between downstream events in the target cells and the upstream morphogen

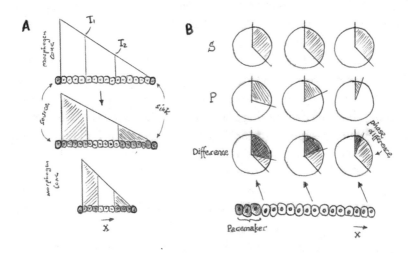

Figure 2: Wolpert's French Flag model versus Goodwin's oscillator-based positional information. (A) In the French Flag model, a morphogen is produced on one end of a tissue (the source) and is degraded at the other (the sink). In the absence of degradation within the tissue, a linear gradient of morphogen concentration will form. The model assumes that cells in the tissue can measure specific concentration thresholds in the gradient (T^1 and T^2). Depending on whether they are above or below a given threshold, cells in the tissue will differentiate in different ways (indicated by different shading). In this way, the morphogen imposes a co-ordinate system on the tissue (called positional information). The model exhibits size regulation due to the law of similar triangles: if the tissue is made smaller, the thresholds remain at the same relative position, and thus preserve proportionality of the pattern (adapted from Wolpert 1968; Wolpert 1969).
(B) Positional information based on phase differences: cells in a tissue oscillate in a co-ordinated manner. This oscillation is called the S-wave. Those cells with the highest oscillation frequency (the pacemaker region) impose their period of oscillation on all other cells in the tissue. Another oscillation originates in the pacemaker region (the P-wave), but travels through the tissue much more slowly than the S-wave. Therefore, cells far away initiate this oscillation later than those close to the pacemaker region, and the phase difference between S- and P-wave increases with distance from the pacemaker region. Cells can measure this difference, which therefore implements positional information analogous to that based on chemical morphogen gradients in the French Flag (Goodwin & Cohen 1969). x indicates distance from the source or pacemaker region.

gradient (Jaeger & Reinitz 2006; Jaeger *et al.* 2008). The third problem concerns size regulation: it only works for linear gradients, but fails for more realistic exponential gradient shapes (Slack 1987).

In contrast, Brian – together with the physicist Morrell H. Cohen – managed to synthesise both temporal and spatial dynamics of pattern formation in an elegant and powerful way. Their model solves all three major problems with Wolpert's French Flag. Goodwin & Cohen (1969) suggested that positional information can be encoded, not in the concentration level of a substance, but in the phase difference of two oscillations propagating through a tissue (Figure 2B). As we have seen above, oscillatory phenomena are common and fundamental in cells, whose basic dynamics are governed by their division cycles (Goodwin 1963). The first of the postulated oscillations, called the synchronising or S-wave, emanates from a source or pacemaker region of the tissue, and induces all cells in the tissue to oscillate at the same frequency (Figure 2B). A second, positional or P-wave, is driven by the S-event in the pacemaker region, but propagates from cell to cell more slowly, creating an increasing phase difference with distance from its source (Figure 2B). Cells in the tissue can sense this phase difference, allowing them to measure their distance from the pacemaker.

In this model, temporal and spatial regulation are tightly coupled to produce a robust and versatile pattern-forming mechanism. It can create much more intricate patterns than the French Flag, including periodically repeating stripes, and hierarchically nested gradient patterns (Figure 2B). In addition, it can achieve size regulation if one considers a third, regulatory or R-wave, triggered every time the phase difference between the S- and P-waves exceeds a certain threshold. This mechanism not only retains proportionality with tissue size, but also buffers the pattern against noise in the period of the oscillations, thereby solving the problem of precise interpretation.

The combination of elegant and simple regulatory principles into hierarchical, complex and robust patterning systems is characteristic of Brian's work. It is more powerful than Wolpert's approach in terms of explaining biological phenomena, such as robustness of patterning or regulative development. On the other hand, it is also much more difficult to understand, since its regulatory principles (oscillations and their phase differences) are much less intuitively accessible than the straightforward implementation of a co-ordinate system by morphogen gradients. Furthermore, oscillations are less easy to detect experimentally than spatial gradients of substances.

It is probably for these reasons that Wolpert's French Flag had a much more immediate impact on experimental research programmes than Brian's oscillator-based model: it provided the motivation for the successful identification and characterisation of a number of candidate morphogen gradients in embryos across a wide range of organisms, from slime moulds to flies to vertebrates (reviewed in Tabata & Takei 2004). A consequence of this was the widespread adoption of a rather static picture of development, centred on stable spatial distributions of morphogens. Yet, as these gradient systems are being studied more closely, and in more quantitative detail, it is becoming increasingly clear that their mechanism of action requires feedback regulation, and is much more complex and dynamic than predicted by the French Flag (Lander 2007; Jaeger *et al.* 2008; Dessaud *et al.* 2008; Wartlick *et al.* 2011), a fact acknowledged by Wolpert himself in recent years (Kerszberg & Wolpert 2007; Wolpert 2011).

At the same time, there is an increasing amount of evidence for the importance of oscillators in developmental processes such as segmentation in centipedes (Chipman *et al.* 2004; Chipman & Akam 2008), insects (Pueyo *et al.* 2008; Sarrazin *et al.* 2012; El-Sherif E *et al.* 2012), and vertebrates (reviewed in Dequéant & Pourquié 2008, see also above). In these pattern-forming systems, travelling waves of oscillatory signals determine the sequential formation of body segments. Moreover, a recent study has shown that such waves can also lead to tissue scaling (Lauschke et al. 2012) as predicted by Brian's model (Goodwin & Cohen 1969). It is far from clear at this point how these waves affect cell differentiation. One model – proposed by Brian's colleague Jonathan Cooke together with the mathematician Christopher Zeeman – was directly inspired by Brian's work on oscillators. It suggests that the oscillations implement a clock, which interacts with a wave front moving through the tissue (Cooke & Zeeman 1976; Cooke 1998). Similar assumptions underlie Brian's own work with one of the co-authors of this chapter, in which temporal oscillations slow down and stop after a characteristic time span to create a periodic spatial pattern (Jaeger & Goodwin 2001; 2002). These models differ quite substantially from Brian's original proposition, but the basic insight that oscillations can lead to robust patterning remains valid.

A theory of form: fields and genes

As we have mentioned at the very beginning of this chapter, Brian strongly criticised the molecular- and gene-centred biology of his time pointing out that biologists had lost sight of the problem of organismic form. Evolutionary theory, for example, in its guise as the Modern Synthetic Theory had basically been transformed into a branch of statistics, dealing with frequency distributions of alleles (specific variants of genes) in populations (see Mayr 2002, for an accessible introduction, or Charlesworth & Charlesworth 2010, for a more technical treatment). Change in these distributions – driven by natural selection – was considered to be the fundamental process of evolution. But natural selection acts on phenotypes: character traits (and behaviours) exhibited by growing or adult organisms. It does not act on genes directly. Yet nobody had (or has) a theory of how such phenotypes arise during development and evolution, and how developmental processes shape and guide the direction evolution can take.

There is a whole, relatively new, discipline called evolutionary developmental biology (evo-devo) that attempts to deal with the issue of how to get from molecular evolution at the level of genes to the phenotypic level (see Wilkins 2001, for an excellent and thorough introduction). However, while some practitioners of evo-devo study evolution at the phenotypic level (see, for example, Minelli 2009), most take a gene-centric, molecular approach (Carroll *et al.* 2004; Davidson 2006; Carroll 2006): they gain their evidence by comparing gene expression, and the effects of its perturbation, across different species. It seems unlikely that this will result in a general theory of organismic form.

Therefore, an alternative, integrative approach is needed, based on the study of complex dynamical systems: we need to understand how the forms of interaction between developmental factors constrain and direct evolution. To achieve this, we first need to understand how developmental processes work. Brian's theory on cellular oscillations and pattern formation (Goodwin & Cohen 1969) is an important first step towards such understanding. But it hardly constitutes a general theory, since it only looks at one particular type of pattern-forming mechanism.

For this reason, Brian went on to develop a more general conceptual framework for the study of organismic form. This framework was first published in Goodwin (1982), and is described in a detailed and systematic manner – including evidence supporting it – in Webster & Goodwin (1996). It is based on the concept of the morphogenetic field, and a scientific discipline called rational morphology, both of which had all but disappeared at the time Brian and Gerry Webster attempted to revive them.

The concept of the morphogenetic field had originally been introduced by Theodor Boveri in 1910. It was further developed by classical embryologists such as Gurwitsch, Weiss, Harrison and Needham, who not only refined its definition, but also provided experimental evidence for field phenomena in development. In fact, fields were the predominant explanatory concept in embryology, until their sudden demise and eclipse by the gene around the fifties (see Gilbert *et al.* 1996, for an historical review).

But what is a morphogenetic field? The problem was that until very recently nobody had a clear answer to this question in terms of molecular mechanisms. Fields tend to be abstract and elusive (just think of electro-magnetic fields in physics, which are utterly mysterious to a large number of people). In fact, Wolpert's French Flag was an attempt at clarifying the definition of a developmental field, although it resulted in a rather minimalistic and static version of the concept (Wolpert 1968; 1969). Brian has several definitions of a morphogenetic field (which he sometimes calls 'generative field' or 'generative mechanism'). A simple, but rather technical version is given in Goodwin & Trainor (1985), where the authors state that a field can be described by dynamical 'equations defining space-time order in relevant variables' (for instance, concentrations of ions, metabolites and gene products, or phase differences in oscillations). More generally, Brian defines a morphogenetic field as a self-organising whole governing the transformation of biological form: it is a dynamic process resulting in the emergence of increasingly complex spatial patterns as development proceeds (Goodwin 1982; Webster & Goodwin 1996). Over and over again, he emphasises that form is dynamic, and the study of morphogenesis must be a study of process.

Brian's definitions of fields definitely improve on earlier attempts in terms of clarity and specificity, but remain abstract and a bit diffuse.

Furthermore, they do not easily connect to specific experiments and research programmes in molecular biology. This definitely contributed to the difficulties in reintroducing the field concept to the mainstream of developmental biology, which was increasingly focused on identifying and characterising the molecular and genetic components underlying developmental processes. This led to modes of explanation that were dominated by static pictures that aimed to capture the logic of development. Unsurprisingly, these pictures were incapable of truly representing the fundamental dynamic and integrative aspects of developmental processes.

In Brian's view, morphogenetic fields are the generators of organismic form, and therefore are the foundation of any theory of the generative origin of phenotypes (Goodwin 1982; Webster & Goodwin 1996). In other words, morphogenetic fields, not genes, are the source of agency during development. They drive and determine developmental outcomes (or phenotypes). The parts of the system only exist for and by means of one another. The importance of any part (such as an individual gene) can only be understood in its context. Organisms have functional and structural unity. This view is called biological structuralism, in reference to the philosophy of language of the same name (Goodwin 1990; Webster & Goodwin 1996). It emphasises the structure, or organisation, of a whole biological system over the type and nature of its individual components, and tries to uncover the dynamic principles that govern the directed transformation of biological form over time.

We therefore have to examine regulatory structure and organisation. As we have seen above, regulatory interactions in complex networks can be described by mathematical equations – a view pioneered in the context of development by Brian. Such a set of mathematical equations (one for the dynamics of each component of the system) is called a dynamical system. The equations describe how the state of each component (for instance, its concentration, its spatial distribution, its biochemical activity, or its electric potential) changes over time. If we solve such a system – in the case of complex, nonlinear problems usually by simulating it on a computer – we can learn about the patterns it can produce.

It turns out that even a very complex, nonlinear dynamical system usually only produces a limited set of robust patterns. These patterns are determined by what are called the attractors and basins of attraction of the system (see Strogatz 2000, for an excellent and accessible introduction to this topic). An attractor is a state of the system, at which it does not change any further. In other words, it has reached a steady state or a regularly repeating cycle. A basin of attraction is the set of initial states (also called initial conditions) that eventually end up at the same attractor. These basins of attraction define the morphological potential of the system, that is, the number and types of patterns it can produce. They also explain the discontinuous and constrained nature of phenotypes (this was also proposed, independently of Brian's efforts, in Oster & Alberch 1982).

During development, these potentials collapse into specific morphologies depending on the conditions (external and internal) that are acting upon them. If they collapse early, we get mosaic development where the fate of different cells in an embryo are rigidly determined at early stages. If they collapse late, we get regulative development where cell fate remains flexible and malleable during early development.

Apart from their limited number, there is another interesting aspect of attractors: many of them result in dynamic patterns that resemble each other in specific ways. For example, there are a large number of different dynamical systems that are able to produce a stripe (off-on-off) pattern across a tissue. These systems can be classified into only a handful of stripe-forming mechanisms that show striking similarities not only in the resulting pattern, but also in the dynamics of expression that lead to this final outcome (Cotterell & Sharpe 2010). This opens the prospect of a logical classification of forms, based on the kind of generative processes that produce them. Brian suggested that it may be possible to arrive at something equivalent to the periodic table of elements for development (Goodwin 1982). Such a table of biological forms would be independent of historical contingencies, since it describes the morphological possibilities from which the actual historical process of evolution can select. Evolved organisms correspond to dynamical systems that possess attractors for robust developmental trajectories and successful life cycles.

This brings us to a very old and contentious issue in morphology and evolution: the problem of essentialism and typology. Before Darwin, classification of animals into species (and higher-order taxa) was based on the definition of ideal forms or types for each species, an idea rooted in Platonic idealism (reviewed in Webster & Goodwin 1996). Horses, for example, were seen as imperfect examples of the ideal horse (the horse type). This research philosophy is called essentialism or typology.

Essentialism and typology were strongly refuted by Darwin, who argued that classification should not be based on any abstract type, but rather common descent (Darwin 1859). There is no ideal type, only individuals with all their imperfections are real. Species (and higher-order taxa) are nothing but names, for something that does not really exist. (Later, Ernst Mayr introduced the biological species concept that defined a species in a less arbitrary way, consisting of individuals that are able to interbreed.) This approach is called nominalism.

Today, most evolutionary biologists adopt Darwin's (or more precisely, Mayr's) nominalist concept of species and other taxa. But still, it is very difficult (if not impossible) to avoid typology (Webster & Goodwin 1996). For example, we can easily distinguish horses from donkeys, and consider them separate species although they *can* interbreed. Moreover, we derive common descent from morphological or nowadays DNA-sequence similarity between organisms, criteria which rely on structural arguments. Finally, we can take nominalism *ad absurdum*: if common descent is the only criterion to define a species then our excrements would belong to the same species as we do! Therefore, common descent is neither sufficient nor practical for biological classification.

Brian, together with Gerry Webster, argued that logical classification of form – based on dynamical systems theory – provides a way out of this dilemma (Goodwin 1982; Webster & Goodwin 1996). If we are able to classify organismic forms based on similarities between their generative processes, we have found a rational way of biological classification, a rational taxonomy. This neither depends on common descent – which is difficult to establish – nor essential types – which do not exist. Instead, individuals in a species are seen as variants of a common underlying generative process. Natural kinds are based on equivalent sets of field equations. Just as in the case of the morphogenetic field, this idea is not new, but goes back to the rational morphologists of the eighteenth and nineteenth century (St. Hilaire, Cuvier, Owen) and Goethe among other pre-Darwinian thinkers

(Webster & Goodwin 1996). What *is* new in Brian's version is that rational taxonomy is put on firm mathematical foundations, since it is based on the quantitative comparison of dynamical systems.

This structuralist view of development and evolution has profound implications. First, it provides an explanation of how phenotypes (which are what is being selected) come into being. It shows that surprisingly few biological forms and transitions between them are possible, which in turn explains why each species is qualitatively different from other species since no continuum of forms is allowed (although environmental or genetic influences can generate gradations in forms within types). This imposes obvious constraints on the future direction of evolution, and sheds light on the frequent occurrence of convergence (the evolution of similar character traits in the absence of common descent). Furthermore, it provides a foundation for the concept of character homology – the fact that our arms are structurally equivalent to pectoral fins in fish or chicken wings, but also to our legs – by postulating that homologous structures are generated by equivalent morphogenetic fields (see Goodwin & Trainor 1983 or Webster & Goodwin 1996, Chapter 6, for a more detailed discussion).

But what about genes? Brian went very far in his early work by stating that genes are not really relevant for the study of morphogenetic fields (Goodwin 1982). He explicitly used examples of morphogenetic fields (discussed in detail below) that did not include any genetic factors (Goodwin & Trainor 1980; Goodwin & Trainor 1983; Goodwin & Trainor 1985; Brière & Goodwin 1988). But later he revised and clarified these statements. Already in Goodwin (1982) he notes that genes act by setting particular constraints for generative processes, and in Webster & Goodwin (1996) and Goodwin (1999) he argues that genes *are* important components of morphogenetic fields, since they determine parameters, but not the structure or organisation, of most developmental systems. In other words, how and when genes interact definitely has important consequences on pattern formation, although there is no genetic program predetermining development. In the context of the dynamical systems view discussed above, genes play an important – although not exclusive – role in determining the attractors of a developmental system. However, the specific developmental path followed by an organism depends on many additional non-genetic and environmental factors. Therefore, criticisms claiming that Brian did not consider genes as important at all are clearly exaggerated; in reality, Brian was simply arguing strongly that it was important to accept that genes are not necessarily central to an understanding of the *regulatory dynamics* of developmental process.

The second important role of genes in Brian's theories is their stabilising role in evolution. Obviously, genes are important for heredity. They preserve given developmental parameters, and hence given attractor states from generation to generation. In this way they select from a number of morphological potentials and stabilise a specific developmental process in a given evolutionary lineage. Moreover, gene networks have the ability to 'learn' and adapt, similar to the learning abilities exhibited by neural networks (Webster & Goodwin 1996).

In summary, genes *do* play an important role in development and evolution. But there is no genetic program: genes do not have agency, or cause phenotypes. Instead they select and stabilise developmental processes based on morphogenetic fields, and contribute to adaptation through 'learning'.

Spatial harmonics: cleavage patterns, limbs, and fly embryos

As we have seen in the previous section, the concept of the morphogenetic field is absolutely central to Brian's thinking about development and evolution. It provides the causative framework, in which genes and other factors interact to create organismic form. However, we have also seen that the field concept is rather abstract and difficult to understand for anyone who has not been trained in physics or dynamical systems theory. For this reason, Brian used several examples of specific developmental processes to clarify and illustrate the meaning and explanatory power of morphogenetic fields.

The first of these models described the spatial and temporal sequence of the first cell divisions during embryogenesis (Goodwin & Trainor 1980). These cleavage divisions follow distinct stereotypical patterns in different groups of animals: sea urchins have a radial, amphibians a bilateral, and molluscs a spiral mode of cleavage (Figure 3A; Gilbert 2010). The central idea behind Brian's model, which was inspired by the approach of D'Arcy Wentworth Thompson (Thompson

1917; Goodwin 1999), and was developed in collaboration with the theoretical physicist Lynn Trainor, was the following: starting with a spherical egg (which has the minimum possible surface energy of any solid), they assumed that the plane of each successive cleavage division was defined by the principle that it should increase the embryo's surface energy by the smallest possible amount. As it turns out, the geometrical arrangement of division planes predicted by the model resembles the one observed in radial cleavage (Figure 3A). Asymmetrical divisions, and other modes of cleavage (bilateral, spiral) can be obtained from the model by simple scale transformations along particular axes of the embryo.

The Goodwin-Trainor cleavage model is an excellent illustration of what Brian meant by a morphogenetic field. Surface energy (and hence its minimisation) is a field property of the embryo: it is distributed over the entire surface, and energy values at one location depend on overall surface geometry. Therefore, the minimum-energy states corresponding to cleavage stages are determined by the spatial harmonics of the whole system. No localised gene expression is needed, although Brian acknowledged the role of genes, for example, to determine the chirality – left- or right-windedness – of spiral cleavage. This nicely illustrates how genes can choose from one of several intrinsic morphogenetic potentials, and stabilise the resulting specific sequence of morphogenetic events.

On the other hand, the model suffered from several problems. It was extremely abstract and failed to correlate the minimisation of surface energy to local biochemical and biomechanical processes known to be involved in cell division. Later models by other authors were more successful in this regard (reviewed in Forgacs & Newman 2005, Chapter 2). Furthermore, the model was not truly dynamic, a deficiency it shared with D'Arcy Thompson's original work (Goodwin 1999). Instead, it consisted of a series of steady-state solutions explaining each successive cleavage stage. And finally, its predictions were only approximate. Although a later version of the model improved its accuracy (Goodwin & Lacroix 1984) it was only able to reproduce new (but not existing) planes of cleavage divisions at each stage. For these reasons, the cleavage field model was largely abandoned, even by Brian himself, who did not cite it in his later work (for example, Webster & Goodwin 1996).

Goodwin and Trainor (1983) also created a model of limb patterning in vertebrates, based on the same principles as the cleavage model described above. It was formulated in terms of an abstract potential to initiate chondrogenesis, the formation of cartilage (and later bone), which determines the morphological structure of the adult limb (Figure 3B). Pattern formation was caused by a field mechanism, which depended on the shape of the developing limb, as well as an abstract principle of overall minimisation of the chondrogenic potential of the system. The model was able to reproduce the reduction and increase in digit numbers between different groups of vertebrates (and was even used to simulate the fossil limb of an ancient marine

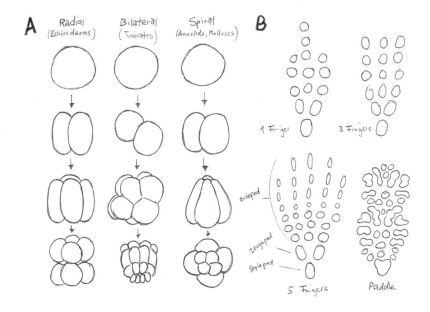

Figure 3: Modes of cleavage and limb patterns.
(A) Different animals show different types of cleavage patterns (the spatial pattern of their first few cell divisions). Here we show the three different types of holoblastic (complete) cleavage that were considered in Goodwin & Trainor (1980). Drawing adapted from Gilbert (2010).
(B) Arrangement of bones – based on centres of chondrogenesis, or cartilage formation – as simulated by the field model in Goodwin & Trainor (1983). The model is able to reproduce patterns with any number of fingers (or toes), as well as the paddle-shaped limb of a fossil marine reptile, Ichthyosaurus, based on variations of the same generative process. The different regions of the vertebrate limb (stylopod, zeugopod, and autopod, roughly corresponding to upper limb, lower limb and hand/foot) are indicated for the 5-fingered limb.

reptile, *Ichthyosaurus*; Figure 3B) in a way which required far fewer assumptions than the alternative view based on Wolpert's concept of positional information (see above and Wolpert 1969).

Just as in the case of the cleavage field model, this conceptual model illustrated the explanatory power of morphogenetic fields in evolution and development. But it was also extremely abstract and did not even attempt to connect to potential cellular or molecular mechanisms in the tissue. Again, Brian abandoned it soon. In his later work on the concept of homology (Webster & Goodwin 1996) he focuses mainly on mechano-chemical models of limb development (Oster *et al.* 1985; Oster *et al.* 1988), which are more specific in terms of cellular and tissue mechanics, but very similar to Brian's model concerning their field-like properties.

As our final example in this section, we describe a series of models of pattern formation and spatial harmonics in the early embryo of the fruit fly *Drosophila melanogaster*, which Brian developed in collaboration with Stuart Kauffman and Axel Hunding (Goodwin & Kauffman 1990; Kauffman & Goodwin 1990; Hunding *et al.* 1990). These models are reviewed in detail in Chapter 7 of Webster & Goodwin (1996), and are discussed in relation to contemporary modelling efforts in Jaeger (2009).

The first of these *Drosophila* models was based on the observation that the expression patterns of segmentation genes (which determine the position of the body segments later in development) go through a cascade of frequency-doubling events. Segmentation genes encode transcription factors that regulate other segmentation genes, forming a hierarchical regulatory network. They can be subdivided into four classes – maternal co-ordinate, gap, pair-rule and segment-polarity genes – corresponding to distinct tiers in the regulatory hierarchy of the segmentation gene network (reviewed in Akam 1987). Maternal co-ordinate and gap genes, which are high in this hierarchy and are activated early, show gradient-like patterns of expression in one to two broad regions along the major (head-to-tail or antero-posterior) axis of the ellipsoid embryo. Pair-rule and segment-polarity genes, which are lower in the hierarchy and become activated later, show periodic patterns of 7 or 14 stripes (Figure 4A). Thus, the number of domains increases as we descend in the hierarchy. Similarly, individual segmentation genes undergo frequency-doubling: the pair-rule gene *even-skipped*, for example, shows an increase from 7 to 14 stripes as its expression pattern matures over time (Figure 4B).

There is no evident biological function for this dynamic behaviour. The developmental relevance of the early expression domain and the late, secondary stripes of *even-skipped* remains unknown. On the other hand, such frequency-doubling events are very common in dynamical systems in transition to deterministic chaos (see, for example, Baker

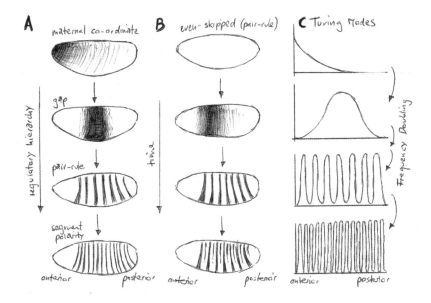

Figure 4: Pattern formation in the early embryo of the fruit fly Drosophila melanogaster.
(A) The segmentation gene hierarchy: the protein products of maternal co-ordinate genes form long-range gradients, which regulate expression of gap genes in one or two broad regions of the embryo. Maternal co-ordinate and gap genes together regulate pair-rule genes, which are expressed in seven stripes. These in turn regulate the expression of segment-polarity genes in fourteen stripes, which form a pre-pattern for segment formation later in development.
(B) Most segmentation genes show increasing numbers of expression domains over time. The pair-rule gene even-skipped is shown as an example. It is not expressed maternally, but shows expression patterns similar to gap, pair-rule and segment-polarity genes at different stages of development. (B) Turing modes are solutions to reaction-diffusion equations that represent a field mechanism of pattern formation. Selection of modes depends on the overall geometry of the embryo, rather than the details of regulatory interactions. We only show those modes which are similar to the expression patterns shown in (A) and (B). Frequency doubling between different modes is indicated. Ellipsoid-shaped embryos are shown with anterior to the left, posterior to the right; dorsal is up, ventral is down.

& Gollub 2008). They are also typically observed between standard solutions (called modes) of Turing reaction-diffusion systems (Figure 4C; Turing 1952). Turing systems are diffusion-driven patterning mechanisms able to produce a large range of spatial patterns from uniform initial conditions (reviewed in Meinhardt 1982; Murray 2002).

Based on this evidence, Brian and Stuart Kauffman created a model in which expression patterns are treated as a symptom, rather than the cause of the observed patterns. They assumed that gene expression is governed by an unknown global field mechanism representing the spatial harmonics of the embryo. The patterns predicted by the model roughly resemble the sequence of expression patterns observed for segmentation genes (Figure 4). But the modelled patterns only depend on the shape and size of the embryo, not the specific interaction between genes. This has the advantage that patterns are robust with respect to genetic perturbations and naturally scale in proportion to embryo size.

Axel Hunding implemented a three-dimensional version of this model, which he simulated on a supercomputer (Hunding *et al.* 1990). This numerical study confirmed the analytical predictions from the earlier model, and reproduced another interesting feature of gene expression that molecular biologists could not explain at the time: the asymmetry (called splay) of segmentation gene expression patterns along the short (back-to-belly or dorso-ventral) axis of the ellipsoid embryo (Figure 4A,B).

The last of Brian's *Drosophila* models, the Four Colour-Wheel Model (Kauffman & Goodwin 1990), is concerned with the interpretation of segmentation gene expression patterns by the cells of the embryo. It aims to explain why these patterns often do not match the regions of the embryo that are affected in segmentation gene mutants, and why we often observe mirror-duplications of entire body regions in such mutants. These phenomena are unexpected and not easily accounted for by the absence of a single factor. To explain them, Brian and Stuart Kauffman proposed that the cells in the embryo do not interpret individual concentrations of regulator proteins, but instead measure the ratio of concentrations between two or three factors (see Jaeger 2009, for a more detailed discussion). This elegant, abstract model yields specific predictions, which have never been tested experimentally in *Drosophila* embryos. They are supported, however, by a recent quantitative study looking at wing development, which found that cells are indeed capable of distinguishing ratios, rather than absolute concentration values of morphogens (Wartlick *et al.* 2011).

The other two models, however – which are the most relevant for our present discussion – were less successful. There was only isolated experimental evidence – based on perturbation of embryos with ether (Ho *et al.* 1987), and later also by magnetic fields (Ho *et al.* 1992) – that suggested the existence of a global field patterning mechanism. On the other hand, even around the time these models were published, there was very convincing experimental evidence that local gene regulatory interactions *are* crucial, and that overall embryo geometry is *not* directly relevant for pattern formation in *Drosophila* (Akam 1989). Brian's beautiful hypothesis was slain by ugly fact (Monk 2000).

Still, these models raised questions that took decades to answer (or even be addressed) by experimental means. They were clearly ahead of their time. Over the last few years, new quantitative experimental techniques have become available that finally enable us to measure gene product concentrations with sufficient accuracy to model patterning robustness and size regulation. As predicted by Brian, these phenomena turn out to be emergent properties of the system, but the molecular mechanism they are based on remains open to debate (see Jaeger 2009, for a review). Similarly, it took twenty years to demonstrate that the splay of segmentation gene expression *does* depend on embryo geometry after all, but not quite as suggested by Brian's model: it is due to the fact that the ventral surface of the embryo (its belly) is more curved than the dorsal side (its back), combined with an asymmetrical distribution of maternal proteins along the dorso-ventral axis (He *et al.* 2010).

Despite their somewhat limited success, it is interesting to consider Brian's field models in their historical context. A certain rate of failure is to be expected under the circumstances: no quantitative experimental approaches, nor suitable quantitative evidence, were available at the time to rigorously test such models. In fact, Brian never claimed that they get the molecular details right. Instead, he intended them to illustrate and clarify a concept – the elusive morphogenetic field. In this abstract aim we believe they succeeded, by illustrating how spatial harmonics *could* explain systems-level properties of the embryo, such as patterning robustness and size regulation. We still lack a mechanistic explanation of these phenomena, but recent quantitative evidence and data-driven modelling studies indicate that they are indeed emergent properties of the regulatory systems underlying pattern formation (Manu *et al.* 2009a,b). Unfortunately, this conceptual aspect of modelling is not really appreciated by many experimental biologists, then or now.

The mechanics of development

Let us now turn to another example. It is Brian's model of morphogenesis and regeneration in the unicellular alga *Acetabularia acetabulum,* once again established through a collaboration with Lynn Trainor. This example stands out for three reasons: first, it describes a generative field based on a detailed molecular mechanism without any genetic component, proving that (at least in principle) genes are not essential for morphogenesis. Second, it has stood the test of time much better than the models described in the previous section, although some details of the underlying mechanism may be incompatible with experimental evidence (see Holloway 2010 or Harrison 2010, Chapter 3, for critical reviews). And third, it can be seen as a progenitor for a number of very successful computational models that examine the interaction of tissue mechanics, growth and chemical processes in plant development.

Acetabularia, called the mermaid's wineglass, is a very unusual organism. In consists of a gigantic single cell, several centimetres long, with a root-like structure (rhizoid), and a stalk carrying a variable number of whorls, which terminates in a cap-like structure (Figure 5A; Dumais *et al.* 2000). The nucleus is located in the rhizoid. When the terminal cap is removed, it regenerates in a reproducible fashion. Interestingly, regeneration also occurs in cell fragments that lack a nucleus, indicating that gene expression is not required for this morphogenetic process.

Based on earlier experimental evidence (Goodwin & Pateromichelakis 1979; Goodwin *et al.* 1983), Goodwin & Trainor (1985) proposed the following model for cap regeneration in *Acetabularia* (Figure 5B): the two main factors considered are strain – imposed by growth processes, the rigid cell wall, and turgor pressure from the cell's internal vacuole – as well as fluctuating concentrations of calcium ions (Ca^{2+}) in the cortical region of the cytoplasm (the region just below the plasma membrane). These factors are tightly coupled through regulatory feedback. Ca^{2+} occurs in two states, either free, or complexed with Ca^{2+}-binding proteins, which are either cytoplasmic or associated with the cytoskeleton (consisting of actin and microtubule protein filaments). Free Ca^{2+} concentration varies in a reversible manner with the viscoelastic properties of the cytoplasm. Viscosity decreases with increasing Ca^{2+} concentrations, while elasticity depends on Ca^{2+} in a complex, nonlinear way. Local changes in viscosity and elasticity in turn induce strain in the cortical cytoplasm, which leads to an increase of free Ca^{2+}.

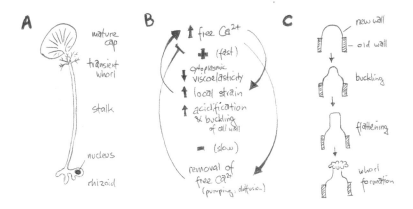

Figure 5. (A) *Acetabularia acetabulum*, the mermaid's wineglass, is a uni-cellular green alga. Its gigantic single cell consists of a rhizoid (anchoring organ), which contains the nucleus, a stalk, transient whorls, and a mature cap, where most of the organism's photosynthetic activity is located.

(B) The Goodwin and Trainor model of tip regeneration in Acetabularia: free Ca^{2+} in the cytoplasm leads to decreased viscosity and elasticity, which leads to increased local strain. This causes acidification and increased flexibility of the overlaying cell wall. Local strain, in turn, leads to the further release of free Ca^{2+}. These interactions form a fast positive feedback of local strain and Ca^{2+} release. At the same time, high free Ca^{2+} concentrations lead to increased downregulation of Ca^{2+} due to the activity of pumps located in the cell membrane, and due to Ca^{2+} ions'. diffusion away from their source. This forms a slow negative feedback loop. This coupling of fast positive and slow negative feedback is typical for excitable media.

(C) Morphogenetic sequence of tip regeneration: when the tip (with its cap) is removed, new cell wall is formed over the wound. High Ca^{2+} (due to high strain) leads to the formation of a regeneration tip through local buckling of the cell wall. This tip flattens after a while since the peak of the strain moves from the centre of the tip to form a peripheral ring (annulus). Local buckling (again, due to high Ca^{2+} and local strain) then leads to the formation of the transient whorls. Adapted from Goodwin & Trainor (1985), and Brière & Goodwin (1988).

The model shows that both strain and Ca^{2+} concentration are highest at the apex of the regenerating tip leading to buckling of the cell wall (Figure 5C). It can also explain the regular arrangement of whorls at a later stage of regeneration (Goodwin & Trainor 1985). These pattern-forming mechanisms are robust towards changes in parameter values (Trainor & Goodwin 1986), local fluctuations in Ca^{2+} concentration (Hart *et al.* 1989), and are also present in a generalised model of calcium-cytogel interactions that includes Ca^{2+} flux or transport across the cell membrane (Brière & Goodwin 1990).

Together with Christian Brière, Brian extended the model to simulate the two-dimensional surface of the regenerating tip (Brière & Goodwin 1988). This model also incorporates tip growth, and the interaction of the cortical cytoplasm with the overlaying cell wall. At the apex of the regenerating tip, this interaction is a simple positive feedback loop (Figure 5B). Increased strain in the cytoplasm leads to the release of protons through the plasma membrane. This in turn results in the acidification of the cell wall, which releases Ca^{2+}, further increasing local buckling and strain. In contrast, Ca^{2+} is actively pumped out of the cytoplasm at the periphery of the tip.

This combination of a short-term positive feedback, with a long-term negative one (Figure 5B,C) is very similar in its dynamics to the mechanism creating cellular oscillations described above. It leads to the local amplification and propagation of a morphogenetic signal (Ca^{2+}, in this case). This type of dynamical system is called an excitable medium.

Apart from revealing this novel layer of feedback regulation, the extended model also explained how the regenerating tip becomes increasingly flattened over time (Figure 5C). As it becomes more and more cylindrical, strain and maximum Ca^{2+} concentration become shifted towards the periphery of the tip, forming a ring around the apex. This is where whorl formation will be induced by a repetition of the morphogenetic events described above.

Apart from the fact that neither Ca^{2+} nor strain are directly controlled by gene regulatory events, this system also beautifully illustrates the close coupling of mechanical and chemical mechanisms in a morphogenetic field. Few other models of pattern formation have achieved such integration so far. A notable exception is provided by recent studies concerning growth and pattern formation in the shoot apical meristem – the growing tip of the shoot of higher plants that generates the aerial structures of the plant. Although the shoot apical meristem is multi-cellular, its morphogenesis depends on the control of cell wall mechanics, just as in *Acetabularia*. Furthermore, it too exhibits a reproducible pattern of morphogenesis – termed phyllotaxis – that depends on the tight coupling of mechanical and chemical signals (Newell *et al.* 2008).

The mechanisms underlying phyllotaxis have been studied in some detail in the plant *Arabidopsis thaliana*, using a combination of experimentation and computational modelling. Attention has focused on the interplay between the plant hormone auxin, which provides chemical signalling between cells, and the orientation of microtubules in each cell – structural elements that regulate the pattern of deposition of new cell wall material. Auxin is transported between cells to generate concentration patterns that prefigure morphogenesis; in turn, the sub-cellular localisation of the auxin transporters changes dynamically as morphogenesis proceeds, providing feedback between the morphogenesis of the meristem as a whole and local chemical transport. Recent studies have shown that this feedback is provided by local stress in the cell wall, which regulates the localisation and orientation of both microtubules and auxin transporters (Hamant *et al.* 2008; Heisler *et al.* 2010).

The picture that is emerging from studies of morphogenesis in higher plants (Hamant & Traas 2010; Chickarmane *et al.* 2010) is strikingly similar to the model for *Acetabularia* presented in Goodwin & Trainor (1985). And yet, Brian's pioneering work is rarely acknowledged. One reason for this may be that it was simply too far ahead of its time – and of sufficiently powerful experimental techniques – to have a significant impact on the community.

Robustness and evolution

There are two complementary questions we can ask about biological pattern formation. The first one is: what are the mechanisms that create pattern? How does pattern arise? We have been dealing mainly with this question so far. The second question deals with reliability and variability: how does the pattern change from individual to individual, or under variable environmental conditions? Obviously, this second question is of enormous importance to evolutionary biology. Without variation, there is no evolution. Moreover, the way a trait varies has profound consequences on how easily and how rapidly it can evolve.

One particularly striking feature of biological systems is their robustness against mutation and environmental perturbations (reviewed in Wagner 2005). It is usually quite hard to change a developmental process. Many mutations show no phenotypic effects at all, and fruit flies look like fruit flies whether they grow in the Siberian tundra or the tropical rainforest of Brazil. Traditional molecular biology and evolutionary genetics, with their focus on isolated genes instead of integrated systems, have no explanation for this phenomenon. Even in a dynamical systems context, robustness is not self-evident. Morphogenetic fields are complex, and our experience with engineering indicates that the more complicated a system the more sensitive it becomes to failure. Why is this not the case in biology?

Brian – together with Stuart Kauffman, and the mathematical biologist James D. Murray – suggested an explanation for the robustness of biological systems in terms of their regulatory structure which governs their dynamical behaviour. The argument is quite simple and generally applicable (Goodwin *et al.* 1993). Developmental processes (and hence morphogenetic fields) are organised in a hierarchical manner: one pattern transformation leads to another. Early patterning of the embryo gives way to organogenesis – the formation of specific organs in particular locations at particular points in development – and so on. It is precisely this hierarchical organisation that gives embryology its robustness.

We have seen earlier that morphogenetic fields are best described as dynamical systems (Strogatz 2001). The patterns such a system can produce depend crucially on its initial conditions, that is, its state at the beginning of the process. In our context, initial conditions are given by the output of the patterning system preceding the current stage of development. Early embryonic patterning in *Drosophila*, for instance, provides the starting point for the formation of nerves, muscles, and other organs later in development (see, for example, Lawrence 1992).

Now we need to remember another important aspect of dynamical systems: even complex, nonlinear processes are only able of producing a limited number of patterns, due to the typically low number of attractors in the system. In this way, the choice of initial conditions for any patterning process is naturally limited by the attracting states of those stages of development preceding it. These specific initial conditions are likely to fall into a very small number of basins of attraction, severely restricting the number of attractors the system can reach. In less technical language: previous patterning processes severely limit and select the subset of patterns the current stage of development can produce. This corresponds exactly to the definition of robustness: even if initial conditions (determined by genetics and the environment) are extremely variable, the number of possible outcomes is limited. The same patterns will be produced no matter what the starting point.

This idea can be illustrated with a simple example: the coupling of a Turing patterning mechanism and cell sorting (see Goodwin *et al.* 1993, for a detailed description). As we have seen when discussing pattern formation in *Drosophila*, Turing systems are diffusion-driven mechanisms capable of producing a number of stereotypical output patterns, which are called modes (reviewed in Murray 2002). Let us now assume a mass of black and white cells, which both adhere to each other with different affinities: black cells stick to each other less strongly than white cells do. A ball of cells will form with white cells in the middle surrounded by black cells (Figure 6). This system can be coupled with a Turing patterning mechanism governing the formation of black and white cells. The first two modes of a Turing system on a sphere are (1) a monotonic gradient, and (2) a pattern with a peak or a trough of black cells in the middle. Let us assume that the latter two patterns (peak or trough in the middle) govern the source of black cells. Due to cell adhesion, the sources of black cells will always come to lie at the poles of the sphere, and never in its middle. This immediately constrains the accessible modes of the Turing system: the pattern with a trough of black cells in the middle is selected over the pattern with a peak (Figure 6). The output of the Turing system has been constrained, and pattern variability is decreased, by the coupling of the two mechanisms. An equivalent argument can be made regarding the coupling of cell wall and cytoplasmic dynamics in *Acetabularia* (Goodwin & Trainor 1985; Brière & Goodwin 1988; Goodwin *et al.* 1993).

It is important to emphasise again that Brian's theory of robustness is based on the dynamical behaviour of developmental processes. In this sense, it is a mechanistic theory, since it explains robustness in terms of generative mechanisms (that is, morphogenetic fields).

The type of robust, hierarchical developmental mechanism that Brian had envisaged has now been implicated in patterning during early *Drosophila* embryogenesis. At this stage of development, nuclei in the fly embryo divide very rapidly (about every 10 to 20 minutes). These divisions are not an essential part of the underlying generative mechanism: identical segmentation gene expression patterns can be observed in models with or without a representation of cell division (Gursky *et al.* 2004). They do, however, contribute to the robustness of patterning in a way exactly equivalent to that proposed by Brian. Each division cycle can be treated as a successive stage in a hierarchical patterning system. And each preceding cycle – by setting specific initial conditions – constrains the subset of attractors that can be reached during the subsequent stage (Gursky *et al.* 2006).

Apart from this, evidence concerning the generative source of robustness remains scarce (see Manu *et al.* 2009a,b, for one of the

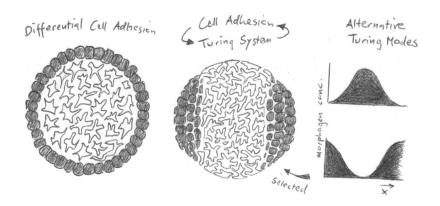

Figure 6. Robust patterning through coupling of cell adhesion and Turing patterning mechanisms. Differential adhesion between black and white populations of cells leads to a pattern where a core of white cells (which strongly adhere to each other) is surrounded by black cells (whose adhesion is weaker). This mechanism can be combined with a Turing pattern mechanism, in which the morphogen induces the differentiation of black cells. This automatically leads to a pattern where the poles of the sphere are formed by black cells. The alternative pattern, where black cells occur in the middle, never occurs because it is excluded by the cell adhesion mechanism. This reduction of alternatives leads to robustness of patterning. Alternative Turing modes are shown to the right. After Goodwin et al. (1993).

very few exceptions, although in this system, robustness is based on regulatory compensation, rather than the hierarchical nature of development). This is quite surprising, given the large amount of interest that this topic has elicited in the last few years. The reason may be that, in contrast to Brian's theory, most current approaches of robustness fall short of providing a mechanistic explanation. Let us illustrate this with an example.

The robust nature of biological systems seems to counteract and contradict their ability to evolve (also called evolvability): how can a robust developmental process be flexible enough to adapt? Recent research by Andreas Wagner and colleagues based on computer simulations of gene regulatory networks has provided an elegant and surprising solution to this paradox (Ciliberti *et al.* 2007a,b; Wagner 2008). This work considers a simple gene network model, in which genes encode transcription factors that control the expression of other genes. Such models can be simulated in large numbers. Each network differs in its 'genotype': the number and type of regulatory interactions they incorporate. Depending on their genotype, different networks will produce distinct characteristic output patterns of genes that are either switched on or off. This output state represents their 'phenotype'.

Such simulations yield some truly surprising results, which do not depend on the details of the model: they confirm earlier propositions based on theoretical considerations (Oster & Alberch 1982; Goodwin 1982) that many different genotypes produce the same phenotype. Moreover, they reveal that many of the genotypes producing the same phenotype are connected by mutation: you can get from one genotype to another by changing only one single interaction in the network. Therefore, such connected genotypes form what is called a neutral network (or more precisely, a neutral network of networks) in genotype space.

Surprisingly, typical neutral networks span a large proportion of genotype space. This implies that a large number of mutations can occur before no further neutral change is possible, and that genotypes producing the same phenotype are not necessarily similar to each other. Both of these implications demonstrate that observed phenotypes are produced in a robust manner. The more connected genotypes produce the same phenotype, the more robust it is.

But how is this compatible with an increased ability to evolve? Let us now consider a non-neutral mutation, one that *does* produce a change in phenotype. Usually, such mutations are detrimental. However, when environmental conditions change, and adaptation becomes a requirement for survival, it is beneficial for an evolving population to be able to produce as much phenotypic diversity as possible. This ability is increased in populations with individuals that have spread out across large neutral networks. The reason for this is as follows: each genotype has a specific subset of phenotypes that are all one mutational step away (its genotypic neighbourhood). The further the distance between two genotypes (in terms of mutational steps) the less likely they are to share the same phenotypes in their neighbourhood. Thus, the farther two individuals are from each other in genotype space, the more likely their offspring will have different phenotypes when mutated, which in turn increases phenotypic variability of the population.

In summary, this theory – which is representative of much current work in the field of network evolution – provides an intuitive resolution to the conflict between robustness and evolvability. Populations of robust organisms are more evolvable because they can explore larger numbers of new phenotypes by mutation. This insight is of absolutely fundamental importance to our understanding of evolutionary dynamics. On the other hand, Wagner's theory leaves robustness itself as an unexplained observation. It does not examine the generative mechanisms that lead to robust phenotypes. Therefore, Brian's theory is complementary to these current efforts, and his contribution to the field remains standing as a challenge for future research.

Conclusions

In this chapter, we have compared Brian's work on biological form and robustness with other approaches to these problems: the molecular branch of evo-devo, or Wagner's theory of robustness, for instance. The situation is very much equivalent in both cases: Brian consistently drew attention to the fact that we lack a general and generative theory that explains the developmental and evolutionary origin of phenotypes, and the robustness with which they are produced. Decades later, we still do. This is a real problem, since we know that the nature of these phenomena greatly influences the rate and direction of evolution. Without such a theory, our view of how life evolved on Earth is simply incomplete, and evolutionary biology remains a discipline based on historical contingency without the capability of understanding (or predicting) the regularities and trends we can observe in evolutionary processes.

We have shown here that Brian's theories attempted to fill this very significant gap in our knowledge. They raised important questions that were (and still are) difficult to tackle. There simply were no adequate experimental approaches, and hence no adequate experimental data, to put his ideas to a rigorous test. It is not surprising then that many of Brian's models and ideas turned out to be incorrect. Brian would not have been surprised or disappointed by this himself. Paraphrasing a soulmate of his, the controversial astronomer Fred Hoyle, he said that it did not matter whether nine of his ideas were wrong as long as the tenth was a really good one. Furthermore, he saw the importance of his models not in precise prediction of experimental results, but rather as conceptual tools that help us think about biological form by clarifying what we mean by morphogenetic fields and generative processes, and by demonstrating their potential implications for morphogenesis and evolution. His field models of cleavage, limb development, and pattern formation in *Drosophila* beautifully illustrate this point.

We believe that Brian's single most important contribution to the fields of cell, developmental and evolutionary biology is his insistence that explanations must be given in terms of generative processes. This is his structuralist theory of biological form, and its astonishing robustness in the face of unpredictable and variable environmental and genetic circumstances (Webster & Goodwin 1996). He argued that, due to this robustness and the discrete nature of phenotypes, it *is* possible to achieve a general theory of biological form, a rational taxonomy, based on the structural similarity of the generative processes that underlie biological pattern formation. He put such a theory on firm mathematical foundations by showing how cellular dynamics and morphogenetic fields can be encoded and studied within the framework of dynamical systems theory. Genes do not determine the outcome of development by themselves, but rather act within this context of dynamic biological structure or organisation. Therefore, Brian *did* provide an alternative to gene-centric biology. Criticisms that see him as negative, and anti-gene, are simply misguided.

Brian's alternative view has long been neglected and misunderstood by experimental biologists. However, it is now staging a comeback in the form of systems biology. The situation in terms of experimental techniques and evidence has changed dramatically in the last few years. Today, we have a whole arsenal of integrative and quantitative methods that allow us to identify all relevant factors involved in almost any given cellular or developmental process, and to measure their changing potentials, activities or concentrations accurately and reliably across space and time. We also have computational and mathematical methods to infer regulatory interactions involved in cellular dynamics or morphogenetic fields from the quantitative datasets we are creating. Systems biology, for the first time, enables us to tackle the questions Brian had raised in a rigorous manner, firmly based on experimental evidence.

We have tried to show how this is happening in fields such as stem cell research, plant morphogenesis, and research of developmental robustness and size regulation in *Drosophila*. Stem cell biologists are now trying to understand cell differentiation in terms of attractors. The parallels between Brian's *Acetabularia* model and modern simulations of the apical shoot meristem are striking, both in their approach and in their insights. Finally, we are only now rediscovering the emergent field-properties of the early *Drosophila* embryo, where dynamical models are used to investigate the astonishing error-correction and size-regulation capabilities of the system. These studies are strongly indebted to Brian's earlier approaches, although they rarely acknowledge him. The questions they ask are exactly the questions that Brian had raised decades ago, and the concepts they use are similar to those used by Brian. It would be fantastic if, in our fast-paced world today, more systems biologists would remember the roots of their field. Brian is, no doubt, one particularly strong branch of that root system.

An Interview with Brian Goodwin: 3

STEPHAN HARDING

Brian on Nature and Evolution

SH: So that initial embeddedness in nature was very important for you because it gave you a sense of the power of nature, and of the importance of life, so that you gradually began to ask yourself the question: 'What is life. How can I begin to understand it?'

BG: Yes, and the first thing as you point out is this sense that that there is some deep power in nature that gives rise to the expression of life, and life is really quite mysterious in all its ramifications. Now I didn't think that life was mysterious – I felt life was something we could figure out.

SH: How old were you when you first started thinking that – teenage or before?

BG: I was [a] teenager. I remember reading a romantic biography of Marie Curie and her research in Paris. What attracted me about that book was the romance of research. There she was looking into nuclear energy and radioactive decay. Figuring out these mysterious aspects of the world that really were very intelligible – very comprehensible that contributed to our scientific understanding. So I articulated this interest in science in terms of research, and I decided that I wanted to do research rather than become a forester because research would obviously appeal to the abstract question that seemed to be part of how you go more deeply into the mysteries of the world.

SH: So looking back to your childhood now, would you say that it gave you an experiential or shamanic way of relating to nature that laid a foundation in your psyche for what was to happen later?

BG: There was some deep experience I had then that led to my rejection of neo-Darwinism – that whole approach to understanding evolution. And so I could only understand this in terms of something that was really – quite deep and mysterious which did not allow me to say: 'Oh yes, that's the explanation – natural selection is the explanation'.

SH: So those experiences when you were young – at eight or nine – gave you a solid core of direct experience in nature that somehow led you to question neo-Darwinism later.

BG: Yes. It's curious that it should take the form of really challenging neo-Darwinism because that was the dominant explanatory theory of evolution at that time. I was brought up on it. But it just stuck in my craw and for some reason I didn't accept it. It took me many years to work out an alternative way of looking at the evolutionary process, which of course has to do with the intrinsic creativity of the natural world. And so I had to struggle with that question for many years.

SH: So you had your deep experiences in nature and also your picking up of your mother's reaction to patriarchy, which folded themselves into your psyche at an early age.

BG: Yes, folded themselves into my psyche and gave me a kind of foundation for rejecting something that was deep in biology which was neo-Darwinism.

SH: When were you first exposed to neo-Darwinism?

BG: Well, you get that (that is, neo-Darwinism) from the first year at university. So I'm talking about when I am eighteen. I've gone to McGill University and I'm exposed to science. I get plenty of biology and the diverse strands to biology that I'm exposed to, one which is evolution, genetics as a dominant strand. But another one is developmental biology. I had an absolutely wonderful teacher in developmental biology, and that turned me on. That was an alternative path for finding an explanation to the phenomenon of evolution, that is, is terms of understanding embryonic development and the emergence of these new forms which is, of course, what speciation is about – the emergence of novelty.

SH: What was the name of your teacher, the one who inspired you to go into developmental biology?

BG: It was Norman Berrell. He was English, originally. He did a lot of work on coelenterates. He spent his summers on the coast collecting organisms for study. So he was an experimentalist, but he was also deeply conceptual. He was somebody who was immersed in Whitehead and the whole idea that everything is an organism.

SH: I see. It sounds like Berrell was a very unconventional academic.

BG: Well, I think he was in the sense that he combined the experimental with the conceptual and the philosophical. He was capable in his lectures of giving you a sense of the coming into being of form. This fourth dimension. Time was tangible and real and he gave you that feeling. He shared that with you.

SH: To what current thought was developmental biology an alternative?

BG: An alternative to natural selection, which focuses on competition and survival of the fitter individuals. What developmental biology focuses on is this intrinsic creativity and coherence of the living realm. You've got a notion of wholeness in the organism, which has to do with its integrity, its health, and its capacity to respond to perturbation, capacity for regulation. Also its capacity to undergo transformation. And so for me the seeds of an alternative explanation of the creativity, coherence and wholeness of life were to be found in developmental biology rather than natural selection. Natural selection is very thin.

SH: So did you see biology as being mainly descriptive and as merely cataloguing various forms of life?

BG: Well, descriptive, but even though this was a mechanical explanation of the emergence of preferred races, I felt that it didn't have any deep principles of organisation. It just said, well anything goes. You know what happens in evolution was that any functional grouping of entities will do. You don't need any principles – it's not like the periodic table of the elements where you need particular types of relationship in order to achieve a taxonomy, a relationship. And so what you've got is necessary and sufficient conditions for the emergence of different elements. Now in biology you had necessary conditions, but not necessary and sufficient conditions. In other words you didn't have a logic that underlay the structure and form of different organisms. I felt quite deeply and intuitively that somewhere principles of that kind could be found.

SH: I see – principles that would explain how organisms of various kinds could generate their forms?

BG: Yes, that's right.

SH: For Darwin forms are just givens, since for him (and the neo-Darwinists) morphologies just appear, somehow?

BG: For Darwin any form is possible. In other words, nothing is excluded in terms pf possible combinations of forms. I mean, that's an exaggeration, but in general that's the case, because it's all functional. What works is what survives. So it's strictly functional, it has no elements of intrinsic structure to it the way the periodic table does. You know there's only certain ways you can get elements emerging as stable forms. You can get unstable forms, but if you going to have stable forms they have to obey certain principles like certain numbers of electrons in the orbital shells. Those are principles of order and organisation. I thought ' Oh – I like that'.

SH: And so you were getting these ideas from your teacher. Was he explicitly speaking about Whitehead in his lectures, or in private with you? How did you know that he was a Whiteheadian? Did he tell you about Whitehead?

BG: I remember going to the library and finding Whitehead's book, *Process and Reality*, and being totally intrigued by it. I mean, here I was in my early twenties. What the hell could I understand of Whitehead? But I understood something from it. I knew that Berrell was interested in Whitehead because he must have said in one of his lectures: 'If you really want to get into this four dimensional world view, where you experience the world as process, coming into being ...'

SH: The fourth dimension being time?

BG: Time. '... Then you need to get into Whitehead. Whitehead is a very good guide'.

SH: Ah, he told you that?

BG: I don't remember explicitly but he must have said that, in his lectures or when we were talking about it.

SH: So by that time you had already finished your degree in biology?

BG: Well, I did a bachelor's degree first of all, a BSc, and that involved a certain amount of research.

SH: So, after your first degree you went to Oxford to do your Masters, as part of your quest to find meaning and order in the processes of life. What did you discover there about natural selection?

BG: Natural selection doesn't do anything about the coherence of the whole organism and the fact that the organism is an integrated whole; and yet in terms of natural selection what you've got is a collection of parts, you've got a collection of characters and each one is separate and each one is selected separately; and so I felt there was a big discrepancy there.

In recent years, of course, evolution and development have come together and they belong together; they belong together because they're the foundation of the creativity that we see in the world, in nature. But whilst I was at Oxford they were separated and they remained separated for some time. There was a sort of tension between the two because genetics obviously took a particular direction; evolution went in the direction of genetics; that was highly reductionist; and then you have these little hereditary particles that become the determinants of the organism; and if these are the determinants of the developmental process, then of course they'd say this is the answer in developmental biology. But developmental biologists have never accepted that because genes don't explain the phenomenon of coherence of form and transformation in organisms.

SH: So developmental biologists even in the seventies, didn't accept neo-Darwinism?

BG: Well, they were split because the whole genome project was dominating the field at that time; there were those who believed that it was going to reveal some very, very important aspects of developmental processes; and, indeed, it did. But, as it turned out, it certainly did not give the answer to embryonic development.

SH: What was missing for you in the neo-Darwinist view?

BG: The coherence, the organisation and the capacity for creative transformation in organisms, these were things that I just found seriously missing in biology; so that was my quest, that was what I was looking for. And at Oxford I went through mathematics and learned various aspects of it and, of course, I didn't know what I was looking for, but you are looking at the aggregate behaviour of large scale entities and then you find out certain principles of order and organisation, and these would interest me and I would say: 'Oh, my goodness, I wonder if this has got the answer to what we're looking for in biology'. And, of course, over the three years at Oxford, I explored different territories and by the end of that time I was ready to do a PhD. And so I had to decide where was I going to do a PhD and with whom.

Brian Goodwin went on to do his PhD under C.H. Waddington at the University of Edinburgh. After his PhD he did a three year postdoc at MIT, and then moved to Sussex University where he was a Reader in Biology until 1984, whereupon he became a full professor in biology at the Open University in Milton Keynes. On his retirement in 1996, he became a member of the faculty at Schumacher College where he worked up to the time of his death in 2009.

Brian Goodwin 1931 –2009

References

Claudio D. Stern: References

Bryant, P.J., Bryant, S.V., & French, V. (1977) Biological regeneration and pattern formation. *Sci Am*, 237, 66–76, 81.

Cooke, J. (1972a) Properties of the primary organization field in the embryo of *Xenopus laevis*. 3. Retention of polarity in cell groups excised from the region of the early organizer. *J Embryol Exp Morphol*, 28, 47–56.

Cooke, J. (1972b) Properties of the primary organization field in the embryo of *Xenopus laevis*. I. Autonomy of cell behaviour at the site of initial organizer formation. *J Embryol Exp Morphol*, 28, 13–26.

Cooke, J. (1972c) Properties of the primary organization field in the embryo of *Xenopus laevis*. II. Positional information for axial organization in embryos with two head organizers. *J Embryol Exp Morphol*, 28, 27–46.

Cooke, J. (1973a) Properties of the primary organization field in the embryo of *Xenopus laevis*. IV. Pattern formation and regulation following early inhibition of mitosis. *J Embryol Exp Morphol*, 30, 49–62.

Cooke, J. (1973b) Properties of the primary organization field in the embryo of *Xenopus laevis*. V. Regulation after removal of the head organizer, in normal early gastrulae and in those already possessing a second implanted organizer. *J Embryol Exp Morphol*, 30, 283–300.

Cooke, J. (1978) Somite abnormalities caused by short heat shocks to pre-neurula stages of *Xenopus laevis*. *J Embryol Exp Morphol*, 45, 283–294.

Cooke, J., & Elsdale, T. (1980) Somitogenesis in amphibian embryos. III. Effects of ambient temperature and of developmental stage upon pattern abnormalities that follow short temperature shocks. *J Embryol Exp Morphol*, 58, 107–118.

Cooke, J., & Zeeman, E.C. (1976) A clock and wavefront model for control of the number of repeated structures during animal morphogenesis. *Journal of Theoretical Biology*, 58, 455–476.

Elsdale, T., Pearson, M., & Whitehead, M. (1976) Abnormalities in somite segmentation following heat shock to *Xenopus* embryos. *J Embryol Exp Morphol*, 35, 625–635.

Eyal-Giladi, H. (1969) Differentiation potencies of the young chick blastoderm as revealed by different manipulations.

I. Folding experiments and position effects of the culture medium. *J Embryol Exp Morphol*, 21, 177–192.

French, V. (1976) Leg regeneration in the cockroach, *Blatella germanica*. II. Regeneration from a non-congruent tibial graft/host junction. *J Embryol Exp Morphol*, 35, 267–301.

French, V. (1978) Intercalary regeneration around the circumference of the cockroach leg. *J Embryol Exp Morphol*, 47, 53–84.

French, V. (1980) Positional information around the segments of the cockroach leg. *J Embryol Exp Morphol*, 59, 281–313.

French, V., Bryant, P.J., & Bryant, S.V. (1976) Pattern regulation in epimorphic fields. *Science*, 193, 969–981.

Goodwin, B.C. (1963) *Temporal Organization in Cells: a dynamic theory of cellular control processes.* London, Academic Press.

Goodwin, B.C. (1976) *Analytical physiology of cells and developing organisms.* London, Academic Press.

Goodwin, B.C., & Cohen, M.H. (1969) A phase-shift model for the spatial and temporal organization of developing systems. *Journal of Theoretical Biology*, 25, 49–107.

Goodwin, B.C., & Pateromichelakis, S. (1979) The role of electrical fields, ions, and the cortex in the morphogenesis of *Acetabularia*. *Planta*, 145, 427–435.

Goodwin, B.C., & Trainor, L.E.H. (1985) Tip and whorl morphogenesis in *Acetabularia* by calcium-regulated strain fields. *Journal of Theoretical Biology*, 117, 79–106.

Jaffe, L.F., & Stern, C.D. (1979) Strong electrical currents leave the primitive streak of chick embryos. *Science*, 206, 569–571.

Maden, M. (1983) The effect of vitamin A on the regenerating axolotl limb. *J Embryol Exp Morphol*, 77, 273–295.

Maden, M., & Goodwin, B.C. (1980) Experiments on developing limb buds of the axolotl *Ambystoma mexicanum*. *J Embryol Exp Morphol*, 57, 177–187.

Maden, M., & Turner, R.N. (1978) Supernumerary limbs in the axolotl. *Nature*, 273, 232–235.

Morata, G., & Lawrence, P.A. (1975) Control of compartment development by the engrailed gene in Drosophila. *Nature*, 255, 614–617.

Nanjundiah, V. (1974) A differential chemotactic response of slime mould amoebae to regions of the early amphibian embryo. *Exp Cell Res*, 86, 408–411.

New, D.A.T. (1955) A new technique for the cultivation of the chick embryo *in vitro*. *J Embryol exp Morph*, 3, 326–331.

O'Shea, P., Goodwin, B.C., & Ridge, I. (1990) A vibrating electrode analysis of extracellular ion currents in Acetabularia acetabulum. *Journal of Cell Science,* 97, 503–508.

Smith, C., Brill, D., Bownes, M., & Ford, C. (1980) Drosophila nuclei replicate in Xenopus eggs. *J Embryol Exp Morphol* 55, 183–194.

Smith, C., McKune, K., Cox, R., & Ford, C. (1983) Preliminary characterisation of inhibitors of DNA polymerase isolated from *Xenopus laevis* early embryos. *Biochim Biophys Acta* 741, 109–115.

Spratt, N.T. (1946) Formation of the primitive streak in the explanted chick blastoderm marked with carbon particles. *J exp Zool,* 103, 259–304.

Stern, C.D., & Goodwin, B.C. (1977) Waves and periodic events during primitive streak formation in the chick. *J Embryol Exp Morphol,* 41, 15–22.

Turner, R.N. (1981) Probability aspects of supernumerary production in the regenerating limbs of the axolotl, Ambystoma mexicanum. *J Embryol Exp Morphol* 65, 119–126.

Waddington, C.H. (1968) *Towards a Theoretical Biology,* Edinburgh University Press.

Webster, G., & Goodwin, B.C. (1996) *Form and transformation: generative and relational principles in biology,* Cambridge University Press.

Wolpert, L. (1969) Positional information and the spatial pattern of cellular differentiation. *Journal of Theoretical Biology,* 25, 1–47.

Michael Ruse: References

Appel, T.A. (1987) *The Cuvier-Geoffroy Debate: French Biology in the Decades Before Darwin.* New York: Oxford University Press.

Brockman, John (1997) 'A New Science of Qualities. A Talk with Brian Goodwin.' April 29, 1997: http://www.edge.org/3rd_culture/goodwin/goodwin_p3.html

Coleman, W. (1964) *Georges Cuvier Zoologist. A Study in the History of Evolution Theory.* Cambridge, Mass.: Harvard University Press.

Darwin, C. (1859) *On the Origin of Species by Means of Natural Selection, or the Preservation of Favoured Races in the Struggle for Life.* London: John Murray.

Eldredge, N. & Gould S.J. (1972) Punctuated equilibria: an alternative to phyletic gradualism. *Models in Paleobiology.* Editor T.J.M. Schopf, pp.82–115. San Francisco: Freeman, Cooper.

Goodwin, B. (2001) *How the Leopard Changed Its Spots,* second edition. Princeton University Press.

Gould, S.J., & R.C. Lewontin. (1979) The spandrels of San Marco and the Panglossian paradigm: a critique of the adaptationist programme. *Proceedings of the Royal Society of London, Series B: Biological Sciences,* 205: pp.581–98.

Gray, A. (1881) *Structural Botany.* 6th ed. London: Macmillan.

Kant, I. (1951) *Critique of Judgement.* New York: Haffner.

Kauffman, S.A. (1995) *At Home in the Universe: The Search for the Laws of Self-Organization and Complexity.* New York: Oxford University Press.

King, David (1996) An Interview with Professor Brian Goodwin, *GenEthics News,* Issue 11. March/April 1996, pp.6–8: http://ngin.tripod.com/article8.htm

Lennox, J.G. (2001) *Aristotle's Philosophy of Biology.* Cambridge University Press.

Maynard Smith, J. (1969) The status of neo-Darwinism. *Towards a Theoretical Biology.* editor C H Waddington, Edinburgh University Press.

Owen, Rev. R. (1894) *The Life of Richard Owen.* London: John Murray.

Paley, W. (1802, 1819) *Natural Theology (Collected Works: IV).* London: Rivington.

Reydams-Schils, G.J., ed. (2003) *Plato's* Timaeus *as Cultural Icon.* University of Notre Dame Press, Ind.

Richards, R.J. (2003) *The Romantic Conception of Life: Science and Philosophy in the Age of Goethe,* University of Chicago Press.

Ruse, M. (2003) *Darwin and Design: Does Evolution have a Purpose–* Cambridge, Mass.: Harvard University Press.

Ruse, M. (2008) *Charles Darwin.* Oxford: Blackwell.

Russell, E.S. (1916) *Form and Function: A Contribution to the History of Animal Morphology.* London: John Murray (reprinted 1982, University of Chicago Press).

Thompson, D.W. (1917) *On Growth and Form,* Cambridge University Press.

Thompson, D.W. (1948) *On Growth and Form, Second Edition.* Cambridge University Press.

Mae-Wan Ho: References

Baldwin, J.M. (1896) A new factor in evolution. *Am. Nat.* 30, pp.441–451, pp.536–553.

Barthélemy-Madaule, M. (1982) *Lamarck the Mythical Precursor* (trans M.H. Shank), MIT Press, Cambridge, Mass.

REFERENCES

Blumberg, M.S. (2009) *Freaks of Nature: What Anomalies Tell Us About Development and Evolution*, Oxford University Press, Oxford.

Bonner, J.T. ed. (1982) *Evolution and Development*, Springer-Verlag, Berlin.

Brakefield, P.M. (2005) Bringing evo devo to life. *PLoS Biology* 3, pp.1693-95.

Brakefield, P.M. (2006) Evo-devo and constraints on selection. *Trends in Ecology and Evolution* 21, pp.362–367.

Burkhardt, R. (1977) *The Spirit of Systems*, Harvard University Press, Cambridge, Mass.

Butler, Samuel (novelist) 23 January 2009, Wikipedia (en.wikipedia.org).

Carroll, S.B. (2005) *The New Science of Evo Devo, Endless Forms Most Beautiful*, W.W. Norton and C., New York.

Coyne, J.A. (2009) Evolution's challenge to genetics. *Nature* 457, pp.382–383.

Darwin, C. (1859) *On the Origin of Species by Means of Natural Selection or the Preservation of Favoured Races in the Struggle for Life*, John Murray, London.

Douady, S. & Couder, Y. (1992) Phyllotaxis as a physical self-organized growth process. *Physical Review Letters* 1992, 68, pp.2098–101.

Dover, G. A. & Flavell, R.B. eds. (1982) *Genome Evolution*, Academic Press, London.

Eldredge, N. & Gould, S. J. (1972) Punctuated equilibria: an alternative to phyletic gradualism. In *Models in Paleobiology* (T.J.M. Schopf, ed.), Freeman, New York.

Fisher, R.A. (1930) *The Genetical Theory of Natural Selection*. Clarendon Press, Oxford.

Gibson, M.C. (2007) Bicoid by the numbers: quantifying a morphogen gradient. *Cell* 130, pp.14–16.

Gilbert, S.F. (2003) The morphogenesis of evolutionary developmental biology. *Int J Dev Biol*, 47, pp.467–77.

Goldschmidt, R.B. (1940). *The Material Basis of Heredity*, Yale University Press, New Haven.

Goodwin, B.C. (1994/1997) *How the Leopard Changed Its Spots*, Weidenfeld & Nicolson, paperback edn. Orion Books, London.

Goodwin, B.C., Webster, G. & Sibatani, A. eds. (1989) *Dynamic Structures in Biology*, Edinburgh University Press, Edinburgh.

Gottlieb, G. (1998) Normally occurring environmental and behavioural influences on gene activity: from Central Dogma to probabilistic epigenetics. *Psychological Review* 105, pp.792–802.

Gould, S.J. (1977) *Ontogeny and Phylogeny*, Belknap Press, Harvard University, Cambridge, Mass.

Gould, S.J. & Eldredge, N. (1972) *Punctuated equilibria: an alternative to phyletic gradualism*. In *Models in Paleobiology*, ed. Schopf, T.J.M., pp.82-115, Freeman, Cooper & Co., San Francisco.

Gray, R. (1988) Metaphors and methods in evolutionary theory. In *Evolutionary Processes and Metaphors*, M.W. Ho & S.W. Fox, eds., Wiley, London.

Gregor,T, Wieschaus, E.F., McGregor, A.P., Bialek, W. & Tank, D.W. (2007) Stability and nuclear dynbamics of the bicoid morphogen gradient. *Cell* 130, 141-52.

Ho M.W. (1983*) Lamarck the Mythical Precursor* – A book review. *Paleon. Ass. Circ.* 14, 10-11.

Ho M.W. (1984a) Where does biological form come from– *Rivista di Biologia* 77, pp.147-79.

Ho, M.W (1984b) Environment and heredity in development and evolution. In *Beyond neo-Darwinism. An Introduction to the New Evolutionary Paradigm* (M.W. Ho & P.T. Saunders, eds.), pp.267-290, Academic Press, London.

Ho, M.W. (1986) Heredity as process. *Rivista di Biologia* 79, pp.407-447.

Ho, M.W. (1987) Evolution by process, not by consequence: implications of the new molelcular genetics on development and evolution. *Int. J. Comp. Psychol.* 1, pp.3-27.

Ho, M.W. (1988a*)* How rational can rational morphology be? A post-Darwinian rational taxonomy based on the structuralism of process. *Rivista di Biologia* 81, pp.11-55.

Ho, M.W. (1988b) Genetic fitness and natural selection: myth or metaphor. In *Proc. 3rd Schneirla Conference*, Lawrence Erlbaum, New Jersey.

Ho, M.W. (1990) An exercise in rational taxonomy. *J. theor. Biol.* 147, pp.43-57.

Ho, M.W. (1993*) The Rainbow and the Worm: The Physics of Organisms*, World Scientific, Singapore, 2nd ed. 1998, reprinted 1999, 2002, 2003, 2005, 2006; 3rd ed. 2008.

Ho, MW. (1998) *Genetic Engineering Dream or Nightmare: The Brave New World of Bad Science and Big Business*, Third World Network, Gateway Books, Macmillan, Continuum, Penang, Malaysia, Bath, UK, Dublin, Ireland, New York, USA, 1998, 1999, 2007 (reprint with extended Introduction).

Ho, M.W. (2003) *Living with the Fluid Genome*, ISIS/TWN, London/Penang.

Ho, M.W. (2004a) Subverting the genetic text. *Science in Society*, 24, pp.6–10.

Ho, M.W. (2004b) Are ultra-conserved elements indispensable? *Science in Society*, 24, p.5.

Ho, M.W. (2004c) To mutate or not to mutate. *Science in Society*, 24, pp.9–10.

Ho, M.W. (2004d) How to keep in concert. *Science in Society*, 24, 12–13.

Ho, M.W. (2008a) In search of the sublime, significant form in science and art. *Science in Society*, 39, pp.4–11.

Ho, M.W. (2008b) GM is dangerous and futile. *Science in Society*, 40, pp.4–8.

Ho, M.W. (2009a) Development and Evolution Revisited. In *Handbook of Developmental Science, Behavior and Genetics* (K. Hood, C. Halpern, G. Greenberg & R. Lerner, eds.), Blackwell Publishing, New York, 2009.

Ho, M.W. (2009b) Epigenetic inheritance, 'what genes remember'. *Science in Society*, 41, pp.4–5.

Ho, M.W. (2009c) Caring mothers strike fatal blow against genetic determinism. *Science in Society*, 41, pp.6–9.

Ho, M.W. (2009d) From genomics to epigenomics. *Science in Society*, 41, pp.10–12.

Ho, M.W. (2009e) Epigenetic toxicology. *Science in Society*, 41, pp.13–15.

Ho, M.W. (2009f) Rewriting the genetic text in human brain development and evolution. *Science in Society*, 41, pp.16–19.

Ho, M.W. (2009g) Epigenetic inheritance through sperm cells, the Lamarckian dimension in evolution. *Science in Society*, 42, pp.40–42.

Ho, M.W. (2009h) Darwin's pangenesis, the hidden history of genetics and the dangers of GMOs. *Science in Society*, 42, pp.42–45.

Ho, M.W. (2009i) *Living with oxygen. Science in Society*, 43, pp.9–12.

Ho M.W. (2010a) Ten years of the human genome. Reams of data and no progress in sight. *Science in Society*, 48, pp.22–25.

Ho, M.W. (2010b) Celebrating the uses of human genome diversity and dissecting the controversies. *Science in Society*, 48, pp.26–29.

Ho, M.W. (2011) Genes don't generate body patterns. *Science in Society*, 52.

Ho, M.W., Matheson, A., Saunders, P.T., Goodwin, B.C., & Smallcombe, A. (1987) Ether-induced segmentation disturbances in Drosophila melanogaster. *Roux's Arch Dev Biol*, 196, pp.511–524.

Ho, M.W. & Saunders, P.T. (1979) Beyond neo-Darwinism: an epigenetic approach to evolution. *J. theor. Biol.*, 78, pp.673–91.

Ho, M.W & Saunders, P.T. (1982) Epigenetic approach to the evolution of organisms – with notes on its relevance to social and cultural evolution. In *Learning, Development, and Culture* (H.C. Plotkin, ed.), pp.343–361.

Ho, M.W. & Saunders, P.T. eds. (1984) *Beyond neo-Darwinism. Introduction to the New Evolutionary Paradigm*, Academic Press, London.

Ho, M.W. & Saunders, P.T. (1993) Rational taxonomy and the natural system with particular reference to segmentation. *Acta Biotheoretica*, 41, pp.289–304.

Ho, M.W. & Saunders, P.T. (1994) Rational taxonomy and the natural system- segmentation and phyllotaxis. In *Models in Phylogeny Reconstruction* (R.W. Scotland, D.J. Siebert & D.M. Williams, eds.), pp.113–24, The Systematics Association Special Volume 52, Oxford Science, Oxford.

Ho, M.W., Stone, T.A., Jerman, I., Bolton, J., Bolton, H., Goodwin, B.C., Saunders, P.T. & Robertson, F. (1992) Brief exposures to weak static magnetic fields during early embryogenesis cause cuticular pattern abnormalities in *Drosophila* larvae. *Physics in Medicine and Biology*, 37, pp.1171–79.

Ho, M.W, Tucker, C., Keeley, D. & Saunders, P.T. (1983) Effects of sucessive generations of ether treatment on penetrance and expression of the bithorax phenocopy. *J. exp. Zool.* 225, pp.1–12.

Hrdy, Sarah B. (2001) *The past, present, and future of the human family.* The Tanner Lectures on Human Values, University of Utah, 27–28 February, 2001 (http://www.citrona.com/TannerHrdy_02.pdf)

Huang, S. (2008a) The genetic equidistance result of molecular evolution is independent of mutation rates. *J Comp Sci Sys Bio.* 1, pp.92–102, (www.omicsonline.com/ArchiveJCSB/Ft01/JCSB1.092.html)

Huang, S. (2008b) Ancient fossil specimens are genetically more distant to an outgroup than extant sister species are. *Riv Bio.*101, pp.93–108

Huang S. (2009) Inverse relationship between genetic diversity and epigenetic complexity. *Nature Precedings:* doi:10.1038/ripre.2009.1751.2: Posted 13 Jan 2009.

Hunter, P. (2008) What genes remember. *Prospect Magazine*, May, (http://www.prospect-magazine.co.uk/article_details.php?id=10140)

Kawai, M. (1962) On the newly-acquired behaviours of the natural troop of Japanese monkeys on Koshima Island. Abstracts of the papers read in the seventh annual meeting of the Society for Primate Researchs, 22–24 November 1962, Japan Monkey Centre, Inuyama. (www.springerlink.com/content/w85j34568088u82q)

Kondo, S. & Miura, T. Reaction-diffusion model as a framework for understanding biological pattern formation. *Science* 2010, 329, pp.1616–20.

Lamarck, J.B. (1809) *Philosophie Zoologique*, Paris.

Løvtrup, S. (1974) *Epigenetics*, John Wiley & Sons, London.

Lowenstein, J. M. (1986) Molecular phylogenetics. *Ann Rev Earth Planet Sci* 14, pp.71–83.

Margoliash, E. (1963) Primary structure and evolution of cytochrome c *Proc Nat Acad Sci* 50, pp.672–79.

Nusslein-Volhard, C. (2006) *Gradients that organize embryo development. A few crucial molecular signals give rise to chemical gradients that organize the developing embryo*, 28 August 2006, Max Planck Institute for Developmental Biology. (www.eb.tuebingen.mpg.de/departments/3-genetics/christiane-nussslein-volhard/gradients-that-organize-embryo-development)

Palmer, A.R. (2004) Symmetry breaking and the evolution of development. *Science* 306, pp.828–833.

Peel A.D., Chipman, A.D. & Akam, M (2005) Arthropod segmentation: beyond the Drosophila paradigm. *Nature Reviews Genetics*, 2005, 6, pp.905–16.

Rothenfluh, H.S. & Steele, E.J. (1993) Origin and maintenance of germ-line V genes. *Immunology and Cell Biology* 71, pp.227–232.

Sapienza, C. (1990) Parental imprinting of genes. *Scient. Am.* 263, pp.26–32.

Saunders, P.T. (1984) Development and evolution. In *Beyond neo-Darwinism. An Introduction to the New Evolutionary Paradigm* (M-W. Ho & P.T. Saunders, eds.), pp.243–63, Academic Press, London.

Saunders, P.T (1989) Mathematics, structuralism and the formal cause in biology. In *Dynamic Structures in Biology* (B.C. Goodwin, G.C. Webster & A. Sibatani, eds). Edinburgh University Press, Edinburgh, 1989, pp.107–120. (Japanese translation: Yoshioka Shoten, Kyoto, 1991.)

Saunders, P.T. (1990) The epigenetic landscape and evolution. *Biological Journal of the Linnean Society*, 39, 125–34.

Saunders, P.T. (1992a) The organism as a dynamical system. In *Thinking About Biology* (W. Stein & F.J. Varela, eds.), Addison-Wesley, Reading, Mass.

Saunders P.T. ed. (1992b) *Alan Turing's Collected Works: Morphogenesis*, Elsevier, Amsterdam.

Saunders PT. (1993) Alan Turing and biology. *IEEE Annals of the History of Computing* 15, pp.33–36.

Saunders, P.T. & Ho, M.W. (1995) Reliable segmentation by successive bifurcation. *Bull Math. Biol,* 57, pp.539–56.

Steele, E.J. (1979) *Somatic Selection and Adaptive Evolution,* Toronto.

Steele E.J. (2008) Reflections on the state of play in somatic hypermutation. *Molec. Immunol.,* 2008, 45, pp.2723–26.

Thompson, D'Arcy W. (1961) *On Growth and Form,* Abridged Edition, Bonner J.T., ed., Cambridge University Press.

Turing A. (1952) The chemical basis of morphogenesis. *Phil Trans B,* 1952, 237, pp.37–72.

Turing, A. (2009) Wikipedia, Pattern_formation_and_ mathematical_biology (en.wikipedia.org/wiki/Alan_Turing)

Valentine, J.W. (2004) *On the Origin of Phyla,* University of Chicago Press, Chicago.

Veeramachaneni, V., Makalowski, W., Galdzicki, M., Sood, R. & Makalowka, I. (2004) Mammalian overlapping genes: the comparative perspective. *Genome Research,* 14, pp.280–86.

Waddington, C.H. (1957) *The Strategy of the Genes,* Allen & Unwin, London.

Webster, G. & Goodwin, B.C. (1982) The origin of species: a structuralist approach. *J. Soc. Biol. Struct.* 5, pp.15–47.

Wright, S. (1964) 'Evolution, Organic.' In *Encyclopedia Britannica* 8, pp.917–29.

Wright, S. (1969; 1978) *Evolution and the Genetics of Populations,* vols. II and IV, University of Chicago Press, Chicago

Zuckerkandl, E. & Pauling, L.B. (1962) Molecular disease, evolution and genetic heterogeneity. In *Horizons in Biochemistry,* Kasha M. & Pullman B. (eds.), pp.189–225, Academic Press, New York.

Craig Millar and David Lambert: References

Beer, G. (1985) *Darwin's Plots: Evolutionary Narratives in Darwin, George Eliot and Nineteenth-Century Fiction,* Ark Paperbacks, London.

Brady, R.H. (1984) The causal dimension of Goethe's morphology. *J. Social Biol. Struct.* 7: pp.325–344.

Darwin, C. (1842) *The Structure and Distribution of Coral Reefs.* John Murray, London.

Goodwin, B. C. (1982) Biology without Darwinian spectacles. *Biologist,* 29, 108–112.

Goodwin, B.C. (1984) 'Changing from an evolutionary to a generative paradigm in biology.' In Pollard, J.W. *Evolutionary Theory: Paths to the Future.* John Wiley & Sons, Chichester.

Goudge, T.A. (1961) *The Ascent of Life: A Philosophical Study of the Theory of Evolution.* George Allen & Unwin, London.

Gould, S.J. (1986) 'Evolution and the triumph of homology, or why history matters.' *American Scientist* 74(1): 60–69.

Lewes, G.H. (1856) *The Life and Works of Goethe,* Tichnor & Fields, Boston.

Russell, E. (1916) *Form and Function.* John Murray, London.

Sahlins, M. (1985) *Islands of History,* University of Chicago Press, Chicago.

Stent, G.S. (1978) *Paradoxes of Progress.* W.H. Freeman and Company, San Francisco.

Margaret Boden: References

Brière, C., & Goodwin, B.C. (1988) 'Geometry and Dynamics of Tip Morphogenesis in *Acetabularia*,' *Journal of Theoretical Biology,* 131: 461–75.

Goethe, J. von (1790) *An Attempt to Interpret the Metamorphosis of Plants* (1790) and *Tobler's Ode to Nature* (1782), trans. A. Arber, *Chronica Botanica,* Vol. 10, no. 2: 63–126 (Waltham, Mass.: Chronica Botanica Co.), 1946.

Goodwin, B.C. (1974) 'Embryogenesis and Cognition', in W. D. Keidel, W. Handler & M. Spreng (eds.), *Cybernetics and Bionics* (Munich: Oldenbourg), 47–54.

Goodwin, B.C. (1976) *Analytical Physiology of Cells and Developing Organisms* (London: Academic Press).

Goodwin, B.C. (1994) *How the Leopard Changed Its Spots: The Evolution of Complexity* (London: Weidenfeld & Nicolson). A new Preface by the author added in 2nd edn. (Princeton, NJ: Princeton University Press, 2001).

Goodwin, B.C. (2006) 'The Unity of Nature and Culture', *Caduceus,* 69 (Autumn).

Goodwin, B.C., & Brière, C. (1992) 'A Mathematical Model of Cytoskeletal Dynamics and Morphogenesis in *Acetabularia.*' In D. Menzel (ed.), *The Cytoskeleton of the Algae* (Boca Raton, FL: CRC Press): 219-238.

Goodwin, B.C., Kauffman, S. A., & Murray, J. D. (1993) 'Is Morphogenesis an Intrinsically Robust Process?' *Journal of Theoretical Biology* 163: 135-144. Goodwin, B.C., & Saunders, P. T., eds., (1989) *Theoretical Biology: Epigenetic and Evolutionary Order from Complex Systems,* Edinburgh University Press.

Goodwin, B.C., & Trainor, L.E.H. (1985) 'Tip and Whorl Morphogenesis in *Acetabularia* by Calcium-regulated Strain Fields', *Journal of Theoretical Biology*, 117: 79-106.

Helmholtz, H. von (1853) 'On Goethe's Scientific Researches', trans H.W. Eve, in H. Helmholtz, *Popular Lectures on Scientific Subjects*, new edn. (London: Longmans Green), 1884, 29–52.

Hull, D.L. (1998) 'A Clash of Paradigms or the Sound of One Hand Clapping (A review of Gerry Webster & Brian Goodwin, *Form and Transformation*, Cambridge University Press, 1996),' *Biology and Philosophy*, 13: 587–595.

Jardine, N. (1991) *The Scenes of Inquiry: On the Reality of Questions in the Sciences*, (Oxford: Clarendon Press).

Kauffman, S.A. (1983) 'Developmental Constraints: Internal Factors in Evolution', in B.C. Goodwin, N. Holder & C.G. Wylie (eds.), *Developmental Evolution*, Cambridge University Press, 195–225.

Kauffman, S.A. (1993) *The Origins of Order: Self-Organization and Selection in Evolution*, Oxford University Press.

Nisbet, H.B. (1972) *Goethe and the Scientific Tradition*, Institute of Germanic Studies, University of London.

Medawar, P.B. (1958) 'Postscript: D'Arcy Thompson and Growth and Form', in R.D. Thompson, D'Arcy Wentworth Thompson: *The Scholar-Naturalist, 1860–1948*. (London: Oxford University Press), 219–233.

Merz, J.T. (1904/12) *A History of European Thought in the Nineteenth Century*, 4 vols., (London: Blackwood).

Sherrington, C.S. (1942) *Goethe on Nature and on Science*, Cambridge University Press.

Solé, R., & Goodwin, B.C. (2000) *Signs of Life: How Complexity Pervades Biology* (New York: Basic Books).

Thompson, D. W. (1917/1942) *On Growth and Form*, Cambridge University Press, 2nd edn., expanded, 1942.

Turing, A.M. (1952) 'The Chemical Basis of Morphogenesis', *Philosophical Transactions of the Royal Society*: B, 237: 37-72.

Von Foerster, H. (1950–55) *Cybernetics, Circular Causal, and Feedback Mechanisms in Biological and Social Systems:* Published Transactions of the Sixth, Seventh, Eighth, Ninth, and Tenth Conferences, 5 vols., (New York: Josiah Macy Foundation).

Waddington, C.H. (1940) *Organisers and Genes*, Cambridge University Press.

Waddington, C.H. ed. (1966–1972) *Towards a Theoretical Biology*, 4 vols., Edinburgh University Press.

Webster, G., & Goodwin, B.C. (1996) *Form and Transformation: Generative and Relational Principles in Biology,* Cambridge University Press.

Fritjof Capra: References

Capra, F. (1996) *The Web of Life.* London: HarperCollins.

Capra, F. (2002) *The Hidden Connections.* London: HarperCollins.

Dawkins, R. (1976) *The Selfish Gene.* Oxford: Oxford University Press.

Goodwin, B. (1994) *How the Leopard Changed Its Spots.* New York: Scribner.

Goodwin, B. (1998) *personal communication.*

Kauffman, S. A. (1991) Antichaos and Adaptation. *Scientific American* 265: pp.64–70.

Kauffman, S. A. (1993) *The Origins of Order.* New York: Oxford University Press.

Luisi, P.L. (1996) *Self-Reproduction of Micelles and Vesicles,* in Prigogine, I. & Rice, S.A. (eds.) *Advances in Chemical Physics,* Vol. 92. New York: John Wiley.

Morowitz, H. (1992) *Beginnings of Cellular Life.* New Haven: Yale University Press.

Mosekilde, E., Aracil, J. & Allen, P.M. (1988) Instabilities and chaos in nonlinear dynamic systems, *System Dynamics Review 4:* pp.14–55.

Prigogine, I. & Glansdorff, P. (1971) *Thermodynamic Theory of Structure, Stability and Fluctuations.* New York: Wiley.

Solé, R. & Goodwin, B. (2000) *Signs of Life.* New York: Basic Books.

Stewart, I. (1997, second edn.) *Does God Play Dice?* Harmondsworth, UK: Penguin Books.

Stewart, I. (1998) *Life's Other Secret.* New York: John Wiley.

Philip Franses: References

Ben Jacob, Eshel *et al.* (2004) Bacterial linguistic communication and social intelligence. *Trends in Microbiology,* Vol. 12.

Bortoft, Henri (1996) *The Wholeness of Nature: Goethe's Way toward a Science of Conscious Participation in Nature,* Lindisfarne Books, NY; Floris Books, Edinburgh.

Cancho, Ramon Ferrer-i, & Solé, Ricard (2003) Least effort and the origins of scaling in human language. *Proc. Nat. Acad Sci.* 100, pp.788-791.

Colquhoun, Margaret & Ewald, Axel (1996)
New Eyes for Plants, Hawthorn Press.

Favareau, D. (2006) The evolutionary history of biosemiotics. In *Introduction to Biosemiotics: The New Biological Synthesis.* Marcello Barbieri (Ed.) Berlin: Springer. pp.1–67.

Franses, Philip (2006) *Living Ambiguity,* Schumacher College MSc Thesis.

Franses, Philip, Hindmarch, Charles, Goodwin, Brian & Murphy, David (2007) *The Language of Living Processes.* (In revision.)

Goethe, Johann Wolfgang von (1817) trans. W.H. Auden & Elizabeth Mayer (1992), *Italian Journey 1786-1788,* Penguin Classics.

Goodwin, Brian (1963) *Temporal Organization in Cells: a dynamic theory of cellular control processes,* Academic Press, London.

Goodwin, Brian (1994) *How the Leopard Changed its Spots,* Princeton University Press.

Goodwin, Brian (2007) *Nature's Due,* Floris Books, Edinburgh.

Hindmarch C, Yao S, Beighton G, Paton J and Murphy D (2006) *A comprehensive description of the transcriptome of the hypothalamoneurohypophyseal system in euhydrated and dehydrated rats.* Proc Natl Acad Sci U S A 103: 1609-1614

Hoffmeyer, J. (2008) *Biosemiotics: An Examination into the Signs of Life and the Life of Signs.* Scranton: University of Scranton Press.

Lu, T. *et al.* (2005) Can Zipf's law be adapted to normalize microarrays? *BMC Bioinformatics,* 6: p.37.

Spencer Brown, G (1969) *Laws of Form,* Bohmeier Verlag, Leipzig.

Theissen G. & Saedler, H. (2001) Plant biology: four quartets, *Nature,* 409, 25 January, pp.469–71.

Waddington, C.H. (1972) *Towards a Theoretical Biology,* vol. 4, Edinburgh University Press.

Johannes Jaeger and Nick Monk: References

Akam, M. (1987) The molecular basis for metameric pattern in the *Drosophila* embryo. *Development,* no. 101, pp.1–22.

Akam, M. (1989) Making stripes inelegantly. *Nature,* no. 341, pp.282–83.

Baker, G.L. & Gollub, J.P. (2008) *Chaotic Dynamics: An Introduction,* Cambridge University Press, Cambridge.

Brière, C. & Goodwin, B. (1988) Geometry and Dynamics of Tip Morphogenesis in *Acetabularia*. *Journal of Theoretical Biology*, no. 131, pp.461–75.

Brière, C. & Goodwin, B.C. (1990) Effects of calcium input/output on the stability of a system for calcium-regulated viscoelastic strain fields. *Journal of Mathematical Biology*, no. 28, pp.585–93.

Carroll, S.B. (2006) *Endless Forms Most Beautiful: The New Science of Evo Devo and the Making of the Animal Kingdom*, W. W. Norton & Co, New York.

Carroll, S.B., Grenier, J. & Weatherbee, S. (2004) *From DNA to Diversity: Molecular Genetics and the Evolution of Animal Design*, Blackwell Publishing, Oxford.

Charlesworth, B. & Charlesworth, D. (2010) *Elements of Evolutionary Genetics*, Roberts & Co., Greenwood Village, CO.

Chickarmane, V., Roeder, A.H.K., Tarr, P.T., Cunha, A., Tobin, C. & Meyerowitz, E.M. (2010) Computational Morphodynamics: A Modeling Framework to Understand Plant Growth. *Annual Review of Plant Biology*, no. 61, pp.65–87.

Chipman, A.D. & Akam, M. (2008) The segmentation cascade in the centipede *Strigamia maritima:* Involvement of the Notch pathway and pair-rule gene homologues. *Developmental Biology*, no. 319, pp.160–69.

Chipman, A.D., Arthur, W. & Akam, M. (2004) A Double Segment Periodicity Underlies Segment Generation in Centipede Development. *Current Biology*, no. 14, pp.1250–55.

Ciliberti, S., Martin, O.C. & Wagner, A. (2007a) Innovation and robustness in complex regulatory gene networks. *Proceedings of the National Academy of Sciences of the United States of America*, no. 104, pp.13591–96.

Ciliberti, S., Martin, O.C. & Wagner, A. (2007b) Robustness Can Evolve Gradually in Complex Regulatory Gene Networks with Varying Topology. *PLoS Computational Biology*, no. 3, e15.

Cooke, J. (1998) A gene that resuscitates a theory – somitogenesis and a molecular oscillator. *Trends in Genetics*, no. 14, pp.85–88.

Cooke, J. & Zeeman, E.C. (1976) A Clock and Wavefront Model for Control of the Number of Repeated Structures during Animal Morphogenesis. *Journal of Theoretical Biology*, no. 58, pp.455–76.

Cotterell, J. & Sharpe, J. (2010) An atlas of gene regulatory networks reveals multiple three-gene mechanisms for interpreting morphogen gradients. *Molecular Systems Biology*, no. 6, pp.425.

Darwin, C. (1859) *On the Origin of Species by Means of Natural Selection*, John Murray, London.

Davidson, E.H. (2006) *The Regulatory Genome: Gene Regulatory Networks in Development and Evolution,* Academic Press, Burlington, MA.

Dequéant, M.-L. & Pourquié, O. (2008) Segmental patterning of the vertebrate embryonic axis. *Nature Reviews Genetics,* no. 9, pp.370–82.

Dessaud, E., McMahon, A.P. & Briscoe, J. (2008) Pattern formation in the vertebrate neural tube: a sonic hedgehog morphogen-regulated transcriptional network. *Development,* no. 135, pp.2489–503.

Dumais, J., Serikawa, K. & Mandoli, D.F. (2000) *Acetabularia:* A Unicellular Model for Understanding Subcellular Localization and Morphogenesis during Development. *Journal of Plant Growth Regulation,* no. 19, pp.253–64.

El-Sherif E, Averof M, Brown SJ (2012). A segmentation clock operating in blastoderm and germband stages of *Tribolium* development. Development 139: 4341–4346.

Forgacs, G. & Newman, S.A. (2005) *Biological Physics of the Developing Embryo,* Cambridge University Press, Cambridge.

Gatherer, D. (2010) So what do we really mean when we say that systems biology is holistic? *BMC Systems Biology,* no. 4, 22.

Gilbert, S.F. (2010) *Developmental Biology,* Sinauer Associates, Sunderland, MA.

Gilbert, S.F. & Sarkar, S. (2000) Embracing Complexity: Organicism for the 21st Century. *Developmental Dynamics,* no. 219, pp.1–9.

Gilbert, S.F., Opitz, J.M. & Raff, R.A. (1996) Resynthesizing evolutionary and developmental biology. *Developmental Biology,* no. 173, pp.357–72.

Goldbeter, A. (1997) *Biochemical Oscillations and Cellular Rhythms: The Molecular Bases of Periodic and Chaotic Behaviour,* Cambridge University Press, Cambridge.

Goodwin, B.C. (1963) *Temporal Organization in Cells,* Academic Press, New York.

Goodwin, B.C. (1994) *How The Leopard Changed Its Spots,* Weidenfeld & Nicolson, London.

Goodwin, B.C. (1966) An entrainment model for timed enzyme syntheses in bacteria, *Nature,* no. 209, pp.479–81.

Goodwin, B.C. (1982) Development and evolution. *Journal of Theoretical Biology,* no. 97, pp.43–55.

Goodwin, B.C. (1990) Structuralism in biology. *Science Progress,* no. 74, pp.227–44.

Goodwin, B.C. (1999) D'Arcy Thompson and the Problem of Biological Form. In: Chaplain, A.J., Sing, G.D. & McLachlan, J.C. (eds.), *On Growth and Form: Spatio-temporal Pattern Formation in Biology,* Wiley & Sons Ltd., Chichester.

Goodwin, B.C. (2007) *Nature's Due: Healing Our Fragmented Culture,* Floris Books, Edinburgh.

Goodwin, B.C. & Cohen, M.H. (1969) A Phase-shift Model for the Spatial and Temporal Organization of Developing Systems. *Journal of Theoretical Biology,* no. 25, pp.49–107.

Goodwin, B.C. & Pateromichelakis, S. (1979) The Role of Electrical Fields, Ions, and the Cortex in the Morphogenesis of *Acetabularia. Planta,* no. 145, pp.427–35.

Goodwin, B.C. & Trainor, L.E.H. (1980) A Field Description of the Cleavage Process in Embryogenesis. *Journal of Theoretical Biology,* no. 86, pp.757–70.

Goodwin, B.C. & Trainor, L.E.H. (1983) The ontogeny and phylogeny of the pentadactyl limb. In: Goodwin, B.C., Holder, N. & Wylie, C.C. (eds.), *Development and Evolution,* Cambridge University Press, Cambridge, pp.75–98.

Goodwin, B.C. & Lacroix, N.H.J. (1984) A Further Study of the Holoblastic Cleavage Field. *Journal of Theoretical Biology,* no. 109, pp.41–58.

Goodwin, B.C. & Trainor, L.E.H. (1985) Tip and Whorl Morphogenesis in *Acetabularia* by Calcium-Regulated Strain Fields. *Journal of Theoretical Biology,* no. 117, pp.79–106.

Goodwin, B.C. & Kauffman, S.A. (1990) Spatial Harmonics and Pattern Specification in Early *Drosophila* Development. Part I. Bifurcation Sequences and Gene Expression. *Journal of Theoretical Biology,* no. 144, pp.303–19.

Goodwin, B.C., Skelton, J.L. & Kirk-Bell, S.M. (1983) Control of regeneration and morphogenesis by divalent cations in *Acetabularia mediterranea. Planta,* no. 157, pp.1–7.

Goodwin, B.C., Kauffman, S.A. & Murray, J.D. (1993) Is Morphogenesis an Intrinsically Robust Process? *Journal of Theoretical Biology,* no. 163, pp.135–44.

Gursky, V.V., Kozlov, K.N., Samsonov, A.M. & Reinitz, J. (2006) Cell divisions as a mechanism for selection in stable steady states of multi-stationary gene circuits. *Physica D,* no. 218, pp.70–76.

Gursky, V.V., Jaeger, J., Kozlov, K.N., Reinitz, J. & Samsonov, A.M. (2004) Pattern formation and nuclear divisions are uncoupled in *Drosophila* segmentation: comparison of spatially discrete and continuous models. *Physica D,* no. 197, pp.286–302.

Hamant, O. & Traas, J. (2010) The mechanics behind plant development. *New Phytologist,* no. 185, pp.369–85.

Hamant, O., Heisler, M.G., Jönsson, H., Krupinski, P., Uyttewaal, M., Bokov, P., Corson, F., Sahlin, P., Boudaoud, A., Meyerowitz, E.M., Couder, Y. & Traas, J. (2008) Developmental Patterning by Mechanical Signals in *Arabidopsis. Science,* no. 322, pp.1650–55.

Harrison, L.G. (2010) *The Shaping of Life: The Generation of Biological Pattern,* Cambridge University Press, Cambridge, UK.

Hart, T.N., Trainor, L.E.H. & Goodwin, B.C. (1989) Diffusion Effects in Calcium-Regulated Strain Fields. *Journal of Theoretical Biology,* no. 136, pp.327–36.

He, F., Wen, Y., Cheung, D., Deng, J., Lu, L.J., Jiao, R. & Ma, J. (2010) Distance measurements via the morphogen gradient of Bicoid in *Drosophila* embryos. *BMC Developmental Biology,* no. 10, 80.

Heisler, M.G., Hamant, O., Krupinski, P., Uyttewaal, M., Ohno, C., Jönsson, H., Traas, J. & Meyerowitz, E.M. (2010) Alignment between PIN1 Polarity and Microtubule Orientation in the Shoot Apical Meristem Reveals a Tight Coupling between Morphogenesis and Auxin Transport. *PLoS Biology,* no. 8, e1000516.

Ho, M.-W., Matheson, A., Saunders, P.T., Goodwin, B.C. & Smallcombe, A. (1987) Ether-induced segmentation disturbances in *Drosophila melanogaster. Roux's Archives of Developmental Biology,* no. 196, pp.511–21.

Ho, M.-W., Stone, T.A., Jerman, I., Bolton, J., Bolton, H., Goodwin, B.C., Saunders, P.T. & Robertson, F. (1992) Brief exposures to weak static magentic field during early embryogenesis cause cuticular pattern abnormalities in *Drosophila* larvae. *Physics in Medicine and Biology,* no. 37, pp.1171–79.

Hogenesch, J.B. & Ueda, H.R. (2011) Understanding systems-level properties: timely stories from the study of clocks. *Nature Reviews Genetics,* no. 12, pp.407–16.

Holloway, D.M. (2010) The role of chemical dynamics in plant morphogenesis. *Biochemical Society Transactions,* no. 38, pp.645–50.

Hunding, A., Kauffman, S.A. & Goodwin, B.C. (1990) Drosophila Segmentation: Supercomputer Simulation of Prepattern Hierarchy. *Journal of Theoretical Biology,* no. 145, pp.369–84.

Jacob, F. & Monod, J. (1961) Genetic regulatory mechanisms in the synthesis of proteins. *Journal of Molecular Biology,* no. 3, pp.318–56.

Jaeger, J. (2009) Modelling the *Drosophila* embryo. *Molecular BioSystems,* no. 5, pp.1549–68.

Jaeger, J. & Goodwin, B.C. (2001) A Cellular Oscillator Model for Periodic Pattern Formation. *Journal of Theoretical Biology,* no. 213, pp.171–81.

REFERENCES

Jaeger, J. & Goodwin, B.C. (2002) Cellular Oscillators in Animal Segmentation. *Silico Biology,* no. 2, pp.111–23.

Jaeger, J. & Reinitz, J. (2006) On the dynamic nature of positional information. *BioEssays,* no. 28, pp.1102–11.

Jaeger, J., Irons, D. & Monk, N. (2008) Regulative Feedback in Pattern Formation: Towards a General Relativistic Theory of Positional Information. *Development,* no. 135, pp.3175–83.

Kaneko, K. (2009) Chaotic expression dynamics implies pluripotency: when theory and experiment meet. *Biology Direct,* no. 4, 17.

Kaneko, K. (2011) Characterization of stem cells and cancer cells on the basis of gene expression profile stability, plasticity, and robustness. *BioEssays,* no. 33, pp.403–13.

Kaneko, K. & Yomo, T. (1997) Isologous diversification: a theory of cell differentiation. *Bulletin of Mathematical Biology,* no. 59, pp.139–96.

Kauffman, S.A. & Goodwin, B.C. (1990) Spatial Harmonics and Pattern Specification in Early *Drosophila* Development. Part II. The Four Colour Wheels Model. *Journal of Theoretical Biology,* no. 144, pp.321–45.

Keller, E.F. (2000) *The Century of the Gene,* Harvard University Press, Cambridge, MA.

Kerszberg, M. & Wolpert, L. (2007) Specifying Positional Information in the Embryo: Looking Beyond Morphogens. *Cell,* no. 130, pp.205–09.

Kitano, H. (2002) Systems biology: a brief overview. *Science,* no. 295, pp.1662–64.

Kruse, K. & Jülicher, F. (2005) Oscillations in cell biology. *Current Opinion in Cell Biology,* no. 17, pp.20–26.

Lander, A.D. (2007) Morpheus unbound: reimagining the morphogen gradient. *Cell,* no. 128, pp.245–56.

Lauschke, V.M. Tsiairis, C.D., François P., Aulehla, A. (2012) Scaling of embryonic patterning based on phase-gradient encoding. Nature 493: 101–105.

Lawrence, P.A. (1992) *The Making of a Fly: The Genetics of Animal Design.* Wiley-Blackwell, Oxford.

Lewis, J. (2008) From signals to patterns: space, time, and mathematics in developmental biology. *Science,* no. 322, pp.399–403.

Manu, Surkova, S., Spirov, A.V., Gursky, V., Janssens, H., Kim, A.-R., Radulescu, O., Vanario-Alonso, C.E., Sharp, D.H., Samsonova, M. & Reinitz, J. (2009a) Canalization of Gene Expression in the *Drosophila* Blastoderm by Gap Gene Cross Regulation. *PLoS Biology,* no. 7, e1000049.

Manu, Surkova, S., Spirov, A.V., Gursky, V., Janssens, H., Kim, A.-R., Radulescu, O., Vanario-Alonso, C.E., Sharp, D.H., Samsonova,

M. & Reinitz, J. (2009b) Canalization of Gene Expression and Domain Shifts in the *Drosophila* Blastoderm by Dynamical Attractors. *PLoS Computational Biology*, no. 5, e1000303.

Maroto, M. & Monk, N.A.M. eds. (2009) *Cellular Oscillatory Mechanisms.* Landes Bioscience/Springer Science & Business Media, New York.

Mayr, E. (2002) *What Evolution Is: From Theory to Fact.* Weidenfeld & Nicolson, London.

Meinhardt, H. (1982) *Models of Biological Pattern Formation.* Academic Press, London.

Minelli, A. (2009) *Forms of Becoming: The Evolutionary Biology of Development.* Princeton University Press, Princeton, NJ.

Momiji, H. & Monk, N.A.M. (2009) Oscillatory Notch-pathway activity in a delay model of neuronal differentiation. *Physical Review E*, no. 80, 021930.

Monk, N.A.M. (2000) Elegant hypothesis and inelegant fact in developmental biology. *Endeavour*, no. 24, pp.170–73.

Monk, N.A.M. (2003) Oscillatory Expression of Hes1, p53, and NF-κB Driven by Transcriptional Time Delays. *Current Biology*, no. 13, pp.1–20.

Monod, J. & Jacob, F. (1962) General conclusions: Teleonomic mechanisms in cellular metabolism, growth and differentiation. *Cold Spring Harbor Symposia on Quantitative Biology*, no. 26, pp.389–401.

Murray, J.D. (2002) *Mathematical Biology. II: Spatial Models and Biomedical Applications.* Springer, Berlin.

Newell, A.C., Shipman, P.D. & Sun, Z. (2008) Phyllotaxis: Cooperation and competition between mechanical and biochemical processes. *Journal of Theoretical Biology*, no. 251, pp.421–39.

Noble, D. (2008) Genes and causation. *Philosophical Transactions of the Royal Society of London A*, no. 366, pp.3001–15.

Novak, B. & Tyson, J.J. (2008) Design principles of biochemical oscillators. *Nature Reviews Molecular Cell Biology*, no. 9, pp.981–91.

Oster, G. & Alberch, P. (1982) Evolution and bifurcation of developmental programs. *Evolution*, no. 36, pp.444–59.

Oster, G.F., Murray, J.D. & Maini, P.K. (1985) A model for chondrogenic condensations in the developing limb: the role of extracellular matrix and cell tractions. *Journal of Embryology and Experimental Morphology*, no. 89, pp.93–112.

Oster, G.F., Shubin, N., Murray, J.D. & Alberch, P. (1988) Evolution and morphogenetic rules: the shape of the vertebrate limb in ontogeny and phylogeny. *Evolution*, no. 42, pp.862–84.

REFERENCES

Pueyo, J.I., Lanfear, R. & Couso, J.P. (2008) Ancestral Notch-mediated segmentation revealed in the cockroach *Periplaneta americana*. *Proceedings of the National Academy of Sciences of the United States of America,* no. 105, pp.16614–19.

Robert, J.S., Hall, B.K. & Olson, W.M. (2001) Bridging the gap between developmental systems theory and evolutionary developmental biology. *BioEssays,* no. 23, pp.954–62.

Sarrazin, A.F., Peel, A.D., Averof, M. (2012) A Segmentation Clock with Two-Segment Periodicity in Insects. Science 336: 338–341.

Shimojo, H., Ohtsuka, T. & Kageyama, R. (2008) Oscillations in Notch Signaling Regulate Maintenance of Neural Progenitors. *Neuron,* no. 58, pp.52–62.

Slack, J.M.W. (1987) Morphogenetic gradients – past and present. *Trends in Biochemical Sciences,* no. 12. pp.200–04.

Solé, R. & Goodwin, B. (2000) *Signs of Life: How Complexity Pervades Biology.* Basic Books, New York.

Strogatz, S.H. (2000) *Nonlinear Dynamics and Chaos: With Applications to Physics, Biology, Chemistry and Engineering.* Perseus Books, New York.

Strogatz, S.H. (2001) Exploring complex networks. *Nature,* no. 410, pp.268–76.

Tabata, T. & Takei, Y. (2004) Morphogens, their identification and regulation. *Development,* no. 131, pp.703–12.

Thompson, D.W. (1917) *On Growth and Form.* Cambridge University Press, Cambridge.

Trainor, L.E.H. & Goodwin, B.C. (1986) Stability analysis on a set of calcium-regulated viscoelastic equations. *Physica D,* no. 21, pp.137–45.

Turing, A.M. (1952) The chemical basis of morphogenesis. *Transactions of the Royal Society London B,* no. 237, pp.37–72.

Veflingstad, S.R., Plahte, E. & Monk, N.A.M. (2005) Effect of time delay on pattern formation: Competition between homogenisation and patterning. *Physica D,* no. 207, pp.245–71.

Waddington, C.H. (1975) *The Evolution of an Evolutionist.* Cornell University Press, Ithaca, NY.

Wagner, A. (2005) *Robustness and Evolvability in Living Systems.* Princeton University Press, Princeton, NJ.

Wagner, A. (2008) Robustness and evolvability: a paradox resolved. *Proceedings of the Royal Society B,* no. 275, pp.91–100.

Wartlick, O., Mumcu, P., Kicheva, A., Bittig, T., Seum, C., Jülicher, F. & González-Gaitán, M. (2011) Dynamics of Dpp Signaling and Proliferation Control. *Science,* no. 331, pp.1154–59.

Webster, G. & Goodwin, B.C. (1996) *Form and Transformation: Generative and Relational Principles in Biology.* Cambridge University Press, Cambridge.

Wilkins, A.S. (2001) *The Evolution of Developmental Pathways.* Sinauer Associates, Sunderland, MA.

Wolkenhauer, O. (2002) Systems biology: The reincarnation of systems theory applied in biology? *Briefings in Bioinformatics,* no. 2, pp.258–70.

Wolpert, L. (1968) The French Flag Problem: A Contribution to the Discussion on Pattern Development and Regulation. In Waddington, C.H. (ed.), *Towards a Theoretical Biology,* Edinburgh University Press, pp.125–33.

Wolpert, L. (1969) Positional Information and the Spatial Pattern of Cellular Differentiation. *Journal of Theoretical Biology,* no. 25, pp.1–47.

Wolpert, L. (1989) Positional information revisited. *Development (Supplement),* no. 107, pp.3–12.

Wolpert, L. (2011) Positional information and patterning revisited. *Journal of Theoretical Biology,* no. 269, pp.359–65.

Brian C. Goodwin: Selected Works

Books

Goodwin, B.C. (1963) *Temporal Organization In Cells: A Dynamic Theory Of Cellular Control Processes.* Academic Press, London.

Goodwin, B. (1976) *Analytical Physiology Of Cells And Developing Organisms.* Academic Press, Harcourt Brace Jovanovich, London & New York.

Goodwin, Brian C., Shibatani, Atsuhiro, Webster, Gerry (1989) *Dynamic Structures in Biology.* Edinburgh University Press, Edinburgh.

Goodwin, B.C. (1994) *How The Leopard Changed Its Spots: The Evolution Of Complexity.* Scribner, New York.

Webster, G. & Goodwin, B.C. (1996) *Form and Transformation: Generative And Relational Principles In Biology.* Cambridge University Press, Cambridge.

Goodwin, B.C. (2000) *Signs of Life: How Complexity Pervades Biology.* Basic Books, New York.

Goodwin, B.C. (2007) *Nature's Due: Healing Our Fragmented Culture.* Floris Books, Edinburgh.

Papers

Goodwin, B.C. & Waygood, E.R. (1954) 'Enzyme studies of mitochondria from barley seedlings,' Succinoxidase Inactivation by a Lecithinase in Barley Seedlings. *Nature,* v.174, pp.517–518.

Goodwin, B. (1961) *Studies in the general theory of development and evolution.* PhD thesis/dissertation, University of Edinburgh, United Kingdom.

Goodwin, B. (1964) *Oscillatory behaviour in cellular control processes; final report, by Brian C. Goodwin.* Cambridge University Press.

Goodwin, B. (1964) A statistical mechanics of temporal organization in cells. *Symposia of the Society for Experimental Biology,* v.18:301–26, Cambridge University Press, England.

Goodwin, B. & Sizer, I. (1965) Effects of spinal cord and substrate on acetylcholinesterase in chick embryonic skeletal muscle. *Developmental Biology*, v.11:1, pp.136–153, Elsevier, New York & London.

Goodwin, B.C. & Sizer, I. (1965) Histone Regulation of Lactic Dehydrogenase in Embryonic Chick Brain Tissue. *Science 9*, v.148:3667, pp.242–244, AAAS, USA,

Goodwin, B.C. (1967) Biological Control Processes and Time. *Annals of the New York Academy of Sciences*, v.138:2, pp.748–758, New York.

1969

Goodwin, B.C. & Cohen, Morrel H. (1969) A phase-shift model for the spatial and temporal organization of developing systems. *Journal of Theoretical Biology*, Volume 25, Issue 1, October 1969, pp. 49–107.

Goodwin, B. (1969) Synchronization of Escherichia coli in a Chemostat by Periodic Phosphate Feeding. *European Journal of Biochemistry*, v.10:3, pp.511–514.

Goodwin, B. (1969) Control Dynamics of ß-Galactosidase in Relation to the Bacterial Cell Cycle. *European Journal of Biochemistry*, v.10:3, pp.515–522.

1971

Oatley, K. & Goodwin, B.C. (1971) The explanation and investigation of biological rhythms. In *Biological rhythms and human performance* (A73-33154 16-04), pp.1–38, Academic Press, London & New York,

1977

Stern, Claudio D. & Goodwin, B.C. (1977) Waves and periodic events during primitive streak formation in the chick. *Embryol Exp Morphol*, v.41, pp.15–22.

Goodwin, B.C. (1978) A cognitive view of biological process. *Journal of Social and Biological Structures*, v.1:2, pp.117–125.

Goodwin, B.C. (1978) Mechanics, Fields and Statistical Mechanics in Developmental Biology. *Journal of Social and Biological Structures*, v.1:2, pp.117–125.

1979

Goodwin, B. & Pateromichelakis, S. (1979) The role of electrical fields, ions, and the cortex in the morphogenesis of Acetabularia. *PLANTA*, v.145:5, pp.427–435.

1980

Goodwin, B.C. & Trainor, L.E.H. (1980) A field description of the cleavage process in embryogenesis. *Journal of Theoretical Biology*, v.85:4, pp.757–770.

Goodwin, B., Maden, M. (1980) Experiments on developing limb buds of the axolotl *Ambystoma mexicanum*. *Journal Embryol Exp Morphol*, v.57, pp.177–187.

1982

Webster, G. & Goodwin, B. (1982) The origin of species: a structuralist approach. *Journal of Social and Biological Structures*, v.5:1, p.15–47.

Goodwin, B.C. (1982) Biology Without Darwinian Spectacles, *Biologist*, 29, pp.108–112.

Goodwin, B.C. (1982) Development and evolution. Genetic Epistemology and Constructionist Biology. *Revue Internationale de Philosophie*, v.36, pp.527–548.

1983

Goodwin, B.C., Skelton J.L. & Kirk-Bell, S.M. (1983) Control of regeneration and morphogenesis by divalent cations in *Acetabularia mediterranea*. *PLANTA*, v.157:1, pp.1–7.

1985

Goodwin, Brian C. & Trainor, L.E.H. (1985) Tip and whorl morphogenesis in *Acetabularia* by calcium-regulated strain fields. *Journal of Theoretical Biology*, v.117:1, pp.79–106.

Goodwin, B.C. (1985) Problems and paradigms: What are the causes of morphogenesis? *BioEssays*, v.3:1, pp.32–36.

1987

Ho, Mae-Wan, Matheson, A., Saunders, Peter T., Goodwin Brian C. & Smallcombe, Anna (1987) Ether-induced segmentation disturbances in Drosophila melanogaster. *Development Genes and Evolution,* v.196:8, pp.511–521.

Goodwin, B.C. (1987) 'A science of qualities,' in *Quantum Implications: Essays in Honour of David Bohm,* Basil J. Hiley, F. David Peat eds. pp. 328–337.

1989

Goodwin, B.C. (1989) Organisms and minds as dynamic forms. *Leonardo,* v.22:1, pp.27–31.

1990

Hunding, A., Kauffman, Stuart A., Goodwin, B.C. (1990) Drosophila segmentation: Supercomputer simulation of prepattern hierarchy. *Journal of Theoretical Biology,* v.145:3, pp.369–384.

Kauffman, Stuart A., Goodwin, B.C. (1990) Spatial harmonics and pattern specification in early Drosophila development. Part I. Bifurcation sequences and gene expression. *Journal of Theoretical Biology,* v.144:3, pp.303–319.

1992

Gordon, Deborah M., Goodwin, Brian C., Trainor, L.E.H. (1992) A parallel distributed model of the behaviour of ant colonies. *Journal of Theoretical Biology,* v.156:3, pp.293–307.

1993

Goodwin, B.C. (1993) Oscillations and Chaos in Ant Societies, *Journal of Theoretical Biology, v.*161:3, pp.343–357.

Goodwin, B.C., Kauffman, S., Murray J.D. (1993) Is Morphogenesis an Intrinsically Robust Process? *Journal of Theoretical Biology,* v.163:1, pp.135–144.

Goodwin, B.C., Miramontes, O., Solé, R.V. (1993) Collective behaviour of random-activated mobile cellular automata. *Physica D: Nonlinear Phenomena* v.63:1–2, pp.145–160.

1997

Goodwin, B.C. (1997) Temporal Organization and Disorganization in Organisms. *Chronobiology International*, v.14:5, pp.531–536.

2000

Goodwin, B.C. (2000) The life of form. Emergent patterns of morphological transformation. La vie des formes. Motifs émergents de transformation morphologique. *Comptes Rendus de l'Académie des Sciences – Series III – Sciences de la Vie* v.323:1, pp.15–21.

2001

Goodwin, Brian C. & Jaeger, Johannes (2001) A Cellular Oscillator Model for Periodic Pattern Formation. *Journal of Theoretical Biology*, v.213: 2, pp.171–181.

2002

Goodwin, B.C. (2002) Oscillatory behavior in enzymatic control processes. *Advances in Enzyme Regulation*, v.3, 1965, pp.425–428, IN1–IN2, 429–430, IN3–IN6, 431–437.

Goodwin, Brian C., & Jaeger, Johannes (2002) Cellular Oscillators in Animal Segmentation. In *Silico Biology*, v.2:2, pp.111–123.

Index

Acetabularia 13, 15, 27, 30, 116, 135, 181, 186
— *acetabulum* 176
amoeba 65
Analytical Physiology of Cells and Developing Organisms 27
Aristotle 56
attractor 130f, 138, 166, 182
— model 134
auxin 179
axolotl (Mexican newt) 27f
—, limb regeneration of 28

Baldwin, James Mark 75
Bateson, William 105
bathwater 52
Beagle, The 102
Bellairs, Angus 21
—, Ruth 21, 33
Belousov-Zhabotinsky reaction 65
Berrell, Norman 188, 190
Bicoid 83
bifurcations, series of 134
biological
— field theory 51
— form 45
biologists
—, evolutionary 59
—, experimental 186
biology
—, developmental 51, 120, 188
—, molecular 34
— of language 142
—, systems 154, 186

biosemiotics 142
Bird, Adrian 19
blackbird 150
Boolean network 133
Bortoft, Henri 147
Boveri, Theodor 164
Bownes, Mary 18
Brière, Christian 178
Brill, Dave 22
Brockes, Jeremy 24
Brockman, John 69
Brown, Spencer 145
Bryant, Peter 18
—, Susan 21

calcium diffusion 13
Cambrian 'explosion' 76
Canadian pine (*Pinus resinosia*) 87
cancer 53
carbon 49
Castaneda, Carlos 22
cellular dynamics 154
centipedes 162
Charlesworth, Brian 18
—, Deborah 18
chemical soup 138
circadian rhythms 155
cleavage field model 172
cockroach limbs 20
Cohen, Morrell H. 24, 161
Cole, Robin 19
Collett, Janet 18
Colquhoun, Margaret 147
complexity theory 127, 140

220

computer model of language 141
— simulations 136
Cook, Captain James 97
Cooke, Jonathan 18, 25, 33
coral reefs 102
coureurs de bois 38
Crick, Francis 24, 110
Curie, Marie 187
Cuvier, Georges 58
cyclic-AMP 29, 31
cytoplasm 116
cytoskeleton 13, 135

Darwin, Charles 46, 58, 100, 102 108
— *Origin of Species* 13
Dawkins, Richard 137
deoxyribonucleic acid *see* DNA
Dictyostelium 29–31
dissipative structures 132
DNA 48, 72, 74, 85, 92, 110, 127, 155
— base changes 84
— base sequence 88, 167
Douady and Couder 86
double helix 115
Dover, Gabriel 104, 107
Driesch, Hans 96, 105, 136
Drosophila 17f, 80–82, 84, 172, 181, 185f
— embryogenesis 182
— *melanogaster* 172
Durston, Tony 30
dynamical
— behaviour 182
— systems 180

ecosystems 128
edge of chaos 134
Edwards, Lawrence 123
Einstein's theory of general relativity 35
Eldredge, Niles 60
electrodynamical processes in development 82

embryologists 115
emergence 132
empiricism 46
energy and matter 128
Engels, Friedrich 100
entropy 144
Ephestia kuhniella 26
epigenetic
—complexity 90
— inheritance 71f
— landscape 78
— theory 75
essentialism and typology 166
evolution and development 191
excitable medium 178
Eyal-Giladi, Hephzibah 32
Eysenck, Hans 20

Fibonacci, Leonardo 67
— sequence 86f
field solutions 50
Fisher, Ronald 109
Foerster, Heinz von 115
Ford, Chris 18, 21, 33
formalism 55
form of an organism 51
Four Colour-Wheel Model 174
fractals 131
French Flag model 159
French, Vernon 20, 35
functionalism 55
functional
— nature of organisms 58
— networks 128

Galileo 70
García-Bellido, Antonio 19
genes 168
genetic
— code 140, 145, 149
— distances 89
— programmes 49
— redundancy 137
genome project 191
geological processes 102

221

Goethe, Johann Wolfgang von 57,
   61, 69, 96, 105, 108–11, 119, 122,
   147
Goethean
— approach 38
— morphology 113
— phenomenology 123
Goldschmidt, Richard 75
Gould, Stephen Jay 60

Haldane, J.B.S. 17
Hawai'i 97
hearer effort 144
Hegel 97
Helmholtz, Hermann von 108, 112f,
   119
heredity 93
high-speed computers 130
Hindmarch, Charles 149
Hogan, Brigid 19
homeostasis 94, 157
homology 47, 57, 111
horses 167
Hox genes 77
human genome 134
— sequence 93
Hunding, Axel 172, 174

Ichthyosaurus 172
idealist philosophy 119
Ingham, Phil 18
integrated systems 154
Intelligent Design Theory 70
Iolanthe 55

Jacob and Monod 155
Jaffe, Lionel 32
jellyfish 63
Jungle Book, Rudyard Kipling's 61

Kaneko and Yomo 158
Kant, Immanuel 57
Kauffman, Stuart 26, 64, 110, 133,
   172

Kilner, Philip 122
Kimura, Motoo 88
King, David 69
Kirk-Bell, Susan 27

Lamarck, Jean-Baptiste 73
Lawrence, Peter 19
Lehmann, Alan 19
Lewin, Ben 18
Lewontin, Richard 60
Lima-de-Faria, Antonio 104, 106
Limnaea 52
living
— networks 129
— systems 128
logical and historical explanations
   97
Longuet-Higgins, Christopher 23

Mace, Georgina 19
macroevolution 90–92
Maden, Malcolm 27, 33, 35
Malthus, Reverend 100
Marx, Karl 100
mathematical
— language 130
— models 158
mathematics 121
Mayr, Ernst 167
McCulloch, Warren 12
Mean, Golden 67
medusa 64
Mendel, Gregor 74, 108
Mendelian genetics 59
meristem 68, 179, 186
metabolism 127
microevolution 90
Modern Synthetic Theory 163
molecular
— biology 24
— clock hypothesis 88
— composition 50
— evolution 88, 91
Morata, Ginés 19

222

Morowitz, Harold 139
morphogenesis 45, 78, 109, 114, 117
morphogenetic field 52, 164, 180
morphological transformations 111
morphology 50, 135
Murphy, David 149
Murray, James D. 180
mystical/shamanic experiences 40

natural selection 73, 83, 88, 115, 137, 140, 189f
Naturphilosophie 57, 119
Navier-Stokes equation 52
negative
— entropy 144
— feedback 156
neo-Darwinian
— theory 72
— view 191
neo-Darwinism 83, 187
networks, neutral 183
New, Denis 30
new phenotype 79
Newton 47, 70
nominalism 167
non-equilibrium systems 129
nonlinear
— dynamics 129, 137
— equations 129
— phenomena 130
Notophthalmus 21
Nurse, Paul 19

Ohno, Susumu 104, 106
ontogeny 86
Onychophora 77
oscillatory dynamics 157
O'Shea, Paul 27
Owen, Richard 58
oxygen 76

Paley, William 56
Pangloss, Dr 61
paradigm shift 154

Pateromichelakis, Stelios 27
periodic table 51
phase space 130
phenotypes 163
phyletic gradualism 60
phyllotaxis 66, 84, 86, 179
phylogeny 86
pine cone 52, 66
Pitts, Walter 12
plants
— metamorphosis of 111f
— Alpine versions of 148
Plato 55
prebiotic evolution 139
pre-Darwinian morphologists 47
Prigogine, Ilya 132
Process and Reality 190
punctuated
— equilibria 71, 76
— equilibrium 60
Pythagoreans 55

Qualities, A Science of 13
Quincey, Christian de 121

Rajneesh, Bhagwan Sri 26
random genetic mutations 72
rational morphology 114
regular solids 56
RNA microarray data 149
Romantics, English 57

Saint-Hilaire, Etienne Geoffroy 58, 104f
salamanders 103
San Marco, Spandrels of 60
Sang, Jimmy H. 17
Schrödinger's wave equation 49
Schumacher College 124, 141, 147, 192
sea urchin eggs 136
selfish gene 137
self-organisation 114f, 137
Serbelloni, Villa 23, 25, 110

Shall, Sydney 19
Sherrington, Charles 109
sign 146
Smith, Christina 22
—, John Maynard 16f, 23, 35, 59
snail shell 52
spatial harmonics 169
speaker effort 144
Spemann, Hans 25
spherical egg 170
sporulation 151
Steiner, Rudolf 123
Stent, Gunther 97
stress-tensor field 136
structuralism 13, 15, 71
— in biology 11, 84, 115
structuralist
— theory 185
— view 168
sunflower 67
surface energy 170
survival of the fittest 74
systemic thinking 127

teleologist 56
Temporal Organization in Cells 12
tetrapod limb 103
thermodynamics 141f
Thompson, D'Arcy Wentworth 11, 35, 62, 77, 96, 104, 106, 109f, 114f, 170
three-letter genetic code 115
Towards a Theoretical Biology 142
Trainor, Lynn 27, 176
Transcendental Meditation 22
transformation 47, 51
transformational 'tree' 85
Tubularia 25
Turing
—, Alan 77, 86, 109, 114f
— mechanism 152

— patterning mechanism 181
— reaction-diffusion systems 174
Turner, Neil 28

Urpflanze 57

vertebrate limb pattern 46
volcanoes 102
Voltaire's *Candide* 61

Waddington, Conrad Hal 12, 17, 23, 78, 96, 106, 110, 142, 155, 192
Waddington's epigenetic landscape 104
Wagner, Andreas 183
Wallace, Hugh 28
Watson, James 110
Webster, Gerry 18, 20, 35, 107, 122, 167
Weijer, Cornelis (Kees) 29
Weismann, August 74
Whitehead, A.N. 121, 188, 190
Whitehead's philosophy 121
Whittle, Robert 18
whorls 178
— symmetries 117
Whyte, Lancelot Law 104, 106
Willnecker, Lür 26
Winfree, Art 26, 32
Wolpert, Lewis 20, 159
Wright, Chauncey 66

Xenopus 122
— *laevis* 25

Yogi, Maharishi Mahesh 22

Zipf's law 143, 145, 149